AF822700

Studies in Systems, Decision and Control

Volume 664

Series Editor

Janusz Kacprzyk, Systems Research Institute, Polish Academy of Sciences, Warsaw, Poland

Editorial Board

Dmitry A. Novikov, Institute of Control Sciences (Director), Russian Academy of Sciences, Moscow, Russia

Peng Shi, School of Electrical and Mechanical Engineering, The University of Adelaide, Adelaide, SA, Australia

Jinde Cao, School of Mathematics, Southeast University, Nanjing, China

Marios Polycarpou, KIOS Research and Innovation Center of Excellence, University of Cyprus, Nicosia, Cyprus

Witold Pedrycz, Department of Electrical and Computer Engineering, University of Alberta, Edmonton, AB, Canada

Fatima Anwar, Process Engineer, Applied Materials (United States), Santa Clara, CA, USA

The series "Studies in Systems, Decision and Control" (SSDC) covers both new developments and advances, as well as the state of the art, in the various areas of broadly perceived systems, decision making and control–quickly, up to date and with a high quality. The intent is to cover the theory, applications, and perspectives on the state of the art and future developments relevant to systems, decision making, control, complex processes and related areas, as embedded in the fields of engineering, computer science, physics, economics, social and life sciences, as well as the paradigms and methodologies behind them. The series contains monographs, textbooks, lecture notes and edited volumes in systems, decision making and control spanning the areas of Cyber-Physical Systems, Autonomous Systems, Sensor Networks, Control Systems, Energy Systems, Automotive Systems, Biological Systems, Vehicular Networking and Connected Vehicles, Aerospace Systems, Automation, Manufacturing, Smart Grids, Nonlinear Systems, Power Systems, Robotics, Social Systems, Economic Systems and other. Of particular value to both the contributors and the readership are the short publication timeframe and the worldwide distribution and exposure which enable both a wide and rapid dissemination of research output.

Indexed by SCOPUS, DBLP, WTI Frankfurt eG, zbMATH, SCImago.

All books published in the series are submitted for consideration in Web of Science.

Rodrick Wallace

Bounded Rationality and Its Discontents

Information and Control Theory Models of Cognitive Dysfunction

Springer

Rodrick Wallace
Division of Epidemiology, The New York
State Psychiatric Institute
Columbia University
New York, NY, USA

ISSN 2198-4182 ISSN 2198-4190 (electronic)
Studies in Systems, Decision and Control
ISBN 978-3-032-18422-1 ISBN 978-3-032-18423-8 (eBook)
https://doi.org/10.1007/978-3-032-18423-8

© The Editor(s) (if applicable) and The Author(s), under exclusive license to Springer Nature Switzerland AG 2026

This work is subject to copyright. All rights are solely and exclusively licensed by the Publisher, whether the whole or part of the material is concerned, specifically the rights of translation, reprinting, reuse of illustrations, recitation, broadcasting, reproduction on microfilms or in any other physical way, and transmission or information storage and retrieval, electronic adaptation, computer software, or by similar or dissimilar methodology now known or hereafter developed.
The use of general descriptive names, registered names, trademarks, service marks, etc. in this publication does not imply, even in the absence of a specific statement, that such names are exempt from the relevant protective laws and regulations and therefore free for general use.
The publisher, the authors and the editors are safe to assume that the advice and information in this book are believed to be true and accurate at the date of publication. Neither the publisher nor the authors or the editors give a warranty, expressed or implied, with respect to the material contained herein or for any errors or omissions that may have been made. The publisher remains neutral with regard to jurisdictional claims in published maps and institutional affiliations.

This Springer imprint is published by the registered company Springer Nature Switzerland AG
The registered company address is: Gewerbestrasse 11, 6330 Cham, Switzerland

If disposing of this product, please recycle the paper.

Preface

Rate Distortion Control Theory (RDCT) provides a mathematical framework for modeling bounded rationality in economics (and other milieus) by formalizing how agents trade off decision accuracy against information-processing costs. This approach roots bounded rationality (Gershman 2015; Howes 2009; Kahneman 2003; Lewis 2014; Lieder 2019; Simon 1955; Simon 1972) in information theory, where agents act as 'lossy information channels' constrained by limited cognitive resources (Gershman 2020, Sims 2016; Sims 2018, Zaslavsky 2018).

Viale (Viale 2020), in the context of individual mental function, places such 'bounded rationality' within a larger perspective that we find useful to adopt more generally:

> ...[M]ental disorder... differs from bounded ecological rationality because it [does] not allow... adaptive cognitive functions... [via] a rigidity and impermeability to the signals from the environment and the lack of a realistic representation of [those signals]...
>
> [Even healthy] individuals are characterized by limitations of logical rationality... that [nonetheless] allow them to successfully adapt to the environment to solve problems and to learn from their mistakes. ...[A] behavior can be defined as pathological... not because it deviates from normative standards, but because it is unable to successfully interact with the social and physical environment, in other words to have cognitive success. The irrationality of the mental disorder is such that it differs from bounded ecological rationality because it [does] not allow... adaptive cognitive functions...
>
> ...[A] limited and ecological rationality... builds on the 'embodied' dimension of cognition. Without this embodiment, limited individual rationality would not be ecological, in that it would not... interact successfully with the environment.

Here, rather than taking RDCT as generating models 'of' bounded rationality and associated matters, we first reconsider the methodology from the perspective of the asymptotic limit theorems of information and control theories, with the intent of constructing robust statistical tools that are 'about' observed data sets, in the sense that regression models, t-tests, and the like are also 'about' data. That is, we attempt to derive a novel set of statistical methods that can provide rational benchmarks against which to compare and contrast real-world, real-time observations.

Second, following the arguments of (Pielou 1977), the resulting—often speculative—probability models can, in and of themselves, be used in a feedback loop with data to generate new speculation regarding real-world cognition and its failures, suggesting further rounds of the empirical or observational studies that are the only source of new knowledge, as opposed to speculation.

The probability models developed here closely linking cognition to regulation are indeed speculation, but, to reiterate, might well be crafted into tools for data analysis and associated but necessarily limited direct remedial and policy-design purposes. The arduous work of fully developing and validating such tools, however, remains for the future.

Finally—again following (Pielou 1977)—we most explicitly do not hold this effort as defining some over-comprehensive 'Grand-X-Principal' leading to a General Relativity-like treatment of cognition, consciousness, and the like. We are aiming toward a useful, working toolbox for real-world applications.

In biology, the most fundamental 'grand principals' have long been known: the 'Reproduction with Variation and Selection' of Darwin, Wallace et al., the assertions by Maturana and Varela (Maturana 1980) that the living state is cognitive at every scale and level of organization, and by (Dretske 1994; Atlan and Cohen 1998), and others, that cognitive phenomena are constrained by the necessary conditions of the asymptotic limit theorems of information theory.

Think about these points for a long moment.

What we do differs significantly from relatively rigid, first-approximation 'physics' approaches to bounded rationality and cognitive dynamics based on narrow interpretations of methods from statistical mechanics (e.g., (Ortega 2013, Ortega 2015)). Such 'physics' methods lead directly to a fixed parametrization of 'temperature' in terms of an underlying Boltzmann Distribution model defining a 'free energy functional maximization' that, supposedly, determines dynamics.

Here, we take the underlying probability distribution—possibly fat-tailed and without defined first or second moments and subject to further 'tuning'—as the fundamental entity, allowing a more flexible definition of 'cognitive temperature'. The approach involves converting 'free energy' via a Legendre Transform into an entropy-analog whose gradients, in the sense of the Onsager approximation of nonequilibrium thermodynamics, define 'forces' driving rate-of-change in underlying resource rate variates. This redefinition of 'temperature' leads to a rich characterization of cognition rate and other phenomena as functions of those resource rates.

The resulting formalism provides a unique perspective on such matters as the Yerkes-Dodson inverse-U signal transduction pattern of cognition rate with stress, and the ubiquity of highly punctuated 'Fisher-Zero' phase transitions in cognitive process at and across a vast range of scales and levels of organization. Some of these transitions represent catastrophic failure of cognition and its necessarily paired regulation. Others, more routinely, involve such matters as the punctuated 'accession to consciousness' of the Global Workspace (and similar) models of consciousness, via groupoid symmetry-breaking (Wallace 2017, Wallace 2022).

Rampant, unconstrained biology, after all, is not for the faint-of-heart.

Ultimately, however, to paraphrase the Perplexity AI chatbot's analysis in the Appendix to the last chapter, while our formal approach articulates cognitive dynamics with rigorous application of asymptotic limit theorems from information theory and nonequilibrium thermodynamics, the broader research landscape affirms our essential insight: that instability, regulation, and the ubiquity of potential failure are not design flaws, but deep, necessary features of real-world embodied cognition.

The first chapter explores best-case embodied 'artificial intelligence' systems, better described as relatively simple machine cognition, as cultural artifacts fully immersed in, and bearing the scars of, path-dependent historical trajectories. The second takes methods from the first to follow the dynamics of institutional cognition under siolization, while the third refines the models in the study of workplace autonomy, well known to contemporary military organizations as 'mission command.' Chapter four explores extensions of the formalism, and Chapter five uses these extensions to study failure modes of embodied-and-regulated cognition, finding that hallucination and analogs to psychopathology are not bugs but inherent features of 'real world' cognition. The last chapter explores a generalized model of post-traumatic stress disorder (PTSD) that may operate at different scales and levels of organization.

The first three chapters, at the expense of some repetition, are independently readable and well-referenced. The last chapters enter Terra Incognita.

A clever reader can take this material and run with it.

New York, USA Rodrick Wallace

References

Atlan, H., Cohen, I.: Immune information, self-organization, and meaning. Int. Immunol. **10**, 711–717 (1998)

Dretske, F.: The explanatory role of information. Philos. Trans. Royal Soc. A **349**, 59–70 (1994)

Gershman, S.J., Horvitz, E., Tenenbaum, J.: Computational rationality: a converging paradigm for intelligence in brains, minds, and machines. Science **349**, 273278 (2015). https://doi.org/10.1126/science.aac6076

Gershman, S.: Origin of perseveration in the trade-off between reward and complexity. Cognition **204**, 104394 (2020). https://doi.org/10.1016/j.cognition.2020.104394

Howes, A., Lewis, R., Vera, A.: Rational adaptation under task and processing constraints: implications for testing theories of cognition and action. Psychol. Rev. **116**, 717 (2009). https://doi.org/10.1037/a0017187

Kahneman, D.: Maps of bounded rationality: psychology for behavioral economics. Am. Econ. Rev. **93**, 14491475 (2003). https://doi.org/10.1257/000282803322655392

Lewis, R.L., Howes, A., Singh, S.: Computational rationality: linking mechanism and behavior through bounded utility maximization. Topics Cogn. Sci. **6**, 279311 (2014). https://doi.org/10.1111/tops.12086

Lieder, F., Griffiths, T.: Resource-rational analysis: understanding human cognition as the optimal use of limited computational resources. Behav. Brain Sci. **43**, 160 (2019). https://doi.org/10.1017/S0140525X1900061X

Maturana, H., Varela, F.: Autopoiesis and Cognition: The Realization of the Living. Reidel, Boston (1980)

Ortega, P., Braun, D.: Thermodynamics as a theory of decision-making with information-processing costs. Proc. Royal Soc. A **469**, 20120683 (2013). https://doi.org/10.1098/rspa.2012.0683

Ortega, P., Braun, D., Dyer, J., Kim, K., Tishby, N.: Information-theoretic bounded rationality. arXiv:1512.06789v1 [stat.ML] (2015)

Pielou, E.: Mathematical Ecology. Wiley, New York (1977)

Simon, H.: A behavioral model of rational choice. Quant. J. Econ. **69**, 99118 (1955). https://doi.org/10.2307/1884852

Simon, H.: Theories of bounded rationality. In: McGuire, C.B., Radner, R. (eds.) Decision and Organization, pp. 161176. North-Holland Publishing Company, Amsterdam (1972)

Sims, C.: Rate-distortion theory and human perception. Cognition **152**:181198 (2016). https://doi.org/10.1016/j.cognition.2016.03.020

Sims, C.: Efficient coding explains the universal law of generalization in human perception. Science **360**, 652656 (2018). https://doi.org/10.1126/science.aaq1118

Viale, R.: Psychopathological irrationality and bounded rationality: why is autism economically rational? In: Viale, R. (ed.) Handbook of Bounded Rationality. Routledge, London (2020)

Wallace, R.: Computational Psychiatry: A Systems Biology Approach to the Epigenetics of Mental Disorders. Springer, New York (2017)

Wallace, R.: Consciousness, Cognition and Crosstalk: The Evolutionary Exaptation of Nonergodic Groupoid Symmetry-Breaking. Springer, New York (2022)

Zaslavsky, N., Kemp, C., Regier, T., Tishby, N.: Efficient compression in color naming and its evolution. PNAS **115**, 79377942 (2018). https://doi.org/10.1073/pnas.1800521115

Competing Interests The author has no competing interests to declare that are relevant to the content of this manuscript.

Contents

About the Author

Rodrick Wallace is a research scientist in the Division of Epidemiology at the New York State Psychiatric Institute, associated with the Columbia University Department of Psychiatry. He has an undergraduate degree in mathematics and a Ph.D. in physics from Columbia and completed post-doctoral training in the epidemiology of mental disorders at Rutgers. He has worked as a public interest lobbyist, conducting empirical studies of fire service deployment, and received an Investigator Award in Health Policy Research from the Robert Wood Johnson Foundation. In addition to material on public health and public policy, he has authored peer-reviewed studies modeling evolutionary process and heterodox economics, as well as quantitative analyses of institutional and machine cognition. He publishes in the military science literature and received one of the UK MoD RUSI Trench Gascoigne Essay Awards.

Chapter 1
AI as Cultural Artifact

The military environment... requires AI solutions that operate under uncertainty...

—*Borchert (2024)*

Large models... [are] most accurate in situations most commonly found in their training sets and least accurate in situations that were rare in data or entirely new...

—*Farrell et al. (2025)*

Every new device and mode of war carries the virus of its own technical, tactical, operational, strategic, or political negation...

—*Colin S. Gray (1999)*

Make peace you fools! What else can you do?

—*Field Marshal Gerd von Runstedt, July 1, 1944*

1.1 Introduction

Human minds are the basic cultural artifacts. As Richerson and Boyd (2006) put it, 'Culture is as much a part of human biology as the enamel on our teeth'. Indeed, as Wallace (2022a) argues at some length, by virtue of the 100 ms time constant of their conscious behaviors, human minds are necessarily stripped-down versions of more complex if much slower phenomena such as immune (Atlan and Cohen 1998) and institutional (Wallace 2020) cognition that can entertain multiple, simultaneous, interacting, tunable 'global workspace spotlights'. Healthy human consciousness is restricted to a single such.

© The Author(s), under exclusive license to Springer Nature Switzerland AG 2026

R. Wallace, *Bounded Rationality and Its Discontents*, Studies in Systems, Decision and Control 664, https://doi.org/10.1007/978-3-032-18423-8_1

Contemporary Artificial Intelligence systems or entities are, of course, also cultural artifacts, particularly in terms of their massive training sets, often scraped from culturally-channeled Internet and other sources, incorporating the blindspots inherent to those sources. But they are also cultural artifacts by virtue of their algorithms, built by humans immersed in cultural streams with longstanding, path-dependent historical trajectories that will reflect themselves in the architecture of those algorithms, just as those same trajectories are reflected in the architectures of the buildings containing the AI machineries.

Here, we will explore something of the necessary influences of culture on the dynamics of *embodied* AI systems acting under demands that, in human terms, would be experienced as 'stress'. Our embrace of 'embodiment' for AI, in this analysis, represents a key divergence from much contemporary perspective, as outlined in the Appendix and through a long and persisting debate in the literature (e.g., Dreyfus 1972; Dreyfus and Dreyfus 1986; Fjelland 2020; Svensson 2023). This is no small thing as, according to our analysis, insufficiently embodied AI will necessarily be locked in to massive levels of hallucination. Ours is, then, a best-case analysis.

Indeed, Wallace (2023), Chap. 6 argues at length that the most effectively embodied real-time–and hence classically 'intelligent'—artificial systems are to be found in High Frequency Stock Trading, coupling an algorithmic 'forebrain' directing trades to an AI 'emotional backbrain' that predicts market fluctuations, often on nanosecond time scales. After all this, the level of 'intelligence', however, appears roughly equivalent to the Arrow Worm, the top predator among marine and brackish water plankton. For Wall Street, the Pentagon, the PLA, and many others, this is, perhaps surprisingly, not a trivial level of capability. See MacKenzie (2024) for a more complete description of HFT and its context.

On very long time scales, of course, sufficiently embodied institutions—those closely coupled to their embedding environments—can also be said to display their own particular forms of intelligence.

The formal tools for this exploration will be the asymptotic limit theorems of information and control theories, as particularly applied to the dynamics of cognitive cultural artifacts of any form.

We will characterize cognition from the perspective of these theorems, but must first place that characterization within the milieu of cultural anthropology before moving on to probability models related to AI stability.

1.2 Cultural Conditioning of Cognition

A fundamental difference between East Asian and Western perception and reasoning that, somewhat surprisingly, underlies much of what we will do here has been the subject of considerable study. Nisbett et al. (2001), following in a long line of research (Markus and Kitayama 1991; Svensson 2023), review an extensive literature on empirical studies of basic cognitive differences between individuals raised in East Asian and Western cultural heritages, which they characterize, respectively, as 'holistic' and 'analytic'.

They find

- Social organization directs attention to some aspects of the perceptual field at the expense of others.
- What is attended to influences metaphysics.
- Metaphysics guides tacit epistemology, that is, beliefs about the nature of the world and causality.
- Epistemology dictates the development and application of some cognitive processes at the expense of others.
- Social organization can directly affect the plausibility of metaphysical assumptions, such as whether causality should be regarded as residing in the field versus in the object.
- Social organization and social practice can directly influence the development and use of cognitive processes such as dialectical versus logical ones.

They conclude that tools of thought embody a culture's intellectual history, that tools have theories built into them, and that users accept these theories, albeit unknowingly, when they use these tools.

Masuda and Nisbett (2006) argue that research on perception and cognition suggests that whereas East Asians view the world holistically, attending to the entire field and relations among objects, Westerners view the world analytically, focusing on the attributes of salient objects. Compared to Americans, East Asians were more sensitive to contextual changes than to focal object changes, These results suggest that there can be cultural variation in what may seem to be basic perceptual processes.

Similarly, Nisbett and Miyamoto (2005) argue that fundamental perceptual processes are influenced by culture. These findings establish a dynamic relationship between the cultural context and perceptual processes. They suggest that perception can no longer be regarded as consisting of processes that are universal across all people at all times.

Varnum et al. (2010) summarize the cognitive and social orientation differences between 'East' and 'West' in two tables we adapt below.

Cognitive patterns

	Analytic Cognition	Holistic Cognition
Attention	Field Independent Narrow Focus on salient objects with intent to manipulate them	Field Dependent Broad Focus on relationship of elements, background
Categorization	Taxonomic, focus on a single dimension or shared property	Thematic, focus on functional relationship or overall similarity
Attribution	Dispositional Traits and attributes of individuals determine events	Situational External forces, context, & situations determine events
Reasoning	Analytic Use of formal logic Trends continue	Dialectical Middle way philosophy Trend reversals are likely

Social Orientation Patterns

	Independent Social Orientation	Interdependent Social Orientation
Values & Beliefs	Individualism Autonomy	Collectivism Harmony
Self	Independent Self-Construal Personal social identity Self as bounded	Interdependent Self-Construal Relational social identity Self as overlapping with close others
Emotions	Higher propensity of socially disengaging emotions Happiness as a disengaging emotion	Higher propensity of socially engaging emotions Happiness as an engaging emotion
Motivation	Individual Achievement Self-enhancement Ego-inflation	Achievement for in-group Self-criticism Self-other interconnection

Wallace (2007) explores roughly something similar involving culture and inattentional blindness.

Jullien (2004) significantly extends matters, summarizing Chinese views on large-scale strategic enterprise:

> For something to be realized in an effective fashion [in the Chinese view], it must come about as an effect. It is always through a process (which transforms the situation), not through a goal that leads (directly) to action, that one achieves an effect, a result...
>
> Any strategy thus seems, in the end, to come down to simply knowing how to implicate an effect, knowing how to tackle a situation upstream in such a say that the effect flows 'naturally' from it...

The obvious inference is that such culturally-driven basic biological perceptual and other differences must also be expressed at and across various scales and levels of organization, including the design and operation of AI and other machine systems.

Indeed, as Clausewitz put the—basically Western cultural—conundrum (On War 1976),

> [Critical analysis and proof]... is bound to be easy if one restricts oneself to the most immediate [i.e., salient] aims and effects. This may be done quite arbitrarily if one isolates the matter from its setting and studies it only under those conditions. But in war, as in life generally, all parts of the whole are interconnected and thus the effects produced, however small their cause, must influence all subsequent military operations and modify their final outcome to some degree, however slight. In the same way, every means must influence even the ultimate purpose.

1.3 A Review of Methodology

The Asymptotic Limit Theorems

As Dretske (1994) noted some thirty years ago, the asymptotic limit theorems from information and control theories constrain all possible cognitive phenomena, 'semantic' or not (Cover and Thomas 2006; Khinchin 1957; Nair et al. 2007). Constructing statistical tools for the analysis of cognitive dynamics and dysfunctions depends on these theorems, just as the development of regression models, t-tests, and the like was based on the Central Limit Theorem.

The asymptotic theorems of interest to us include:

- The Shannon Coding Theorem that prescribes how to code a message in 'typical' sequences for maximum error-free transmission through a particular channel at a maximum 'channel capacity' (and the 'tuning theorem' variant in which the message formally 'transmits' the channel).
- The Source Coding Theorem that splits a set of sequences emitted by an information source into two subsets, a relatively small one that is 'meaningful'—the sequences are consonant with an underlying 'grammar' and 'syntax'—and a much larger 'meaningless' one 'of measure zero'. Only for stationary, ergodic systems is it possible to characterize 'information' and 'information source uncertainty' in terms of a Shannon entropy-analog. For nonergodic systems, each (infinite) signal sequence has it's own scalar source uncertainty that cannot be expressed in 'Shannon entropy' form (Khinchin 1957). It is here that methods must be abducted from statistical physics, introducing as well equivalence-classes of sequences forming a groupoid, leading to groupoid symmetry-breaking phase transitions characterizing 'cognitive' phase transitions, quite unlike those of 'conventional' physical systems.
- The Rate Distortion Theorem (RDT) and an associated Rate Distortion Control Theory that, in a sense, 'tunes' the channel capacity around a maximum-permissible scalar measure of average distortion (which finesses the Friston 'free energy minimization' approach).
- The Data Rate Theorem (DRT) stating that, for stability of an inherently unstable control system, 'control information' must be provided at a rate greater than the system produces it's own 'topological information'. A version of this is, somewhat

surprisingly, 'easily' derived from a time and 'free energy rate' optimization based on the Hartley-Shannon Theorem (Wallace 2025a, Sect. 1.13).

Further consideration gives:

- Cognition implies choice—of a small set of actions from a larger embedding set of what is possible to the system of interest. Such choice reduces uncertainty in a formal manner, implying the existence of an information source 'dual' to the cognitive process studied, at and across the varied scales and levels of organization of the living state (Atlan and Cohen 1998; Dretske 1994; Maturana and Varela 1980).
- Real-world, real-time cognition is always inherently unstable under the impact of the 'topological information' imposed by embedding environmental 'roadways'. The DRT implies that cognition must always be paired with regulation for system stability and long-term function under selection pressures. Indeed,, it can perhaps be argued that the degree of pairing between cognition and regulation measures a system's 'self awareness', a central feature of many consciousness models.
- We impose an adiabatically, piecewise, stationary (APS) 'Born Oppenheimer' model in which the system of interest remains 'close enough' to stationary during the period(s) of interest so that the asymptotic limit theorems are effectively 'anytime algorithms' (e.g., Zilberstein 1996). Between 'pieces' for which this assumption is sufficiently true, we invoke (groupoid) symmetry-breaking and/or Fisher Zero (Dolan et al. 2001) phase transitions.
- The ergodic restriction can be lifted, at the cost of imposing methods abducted from statistical physics, with dynamics approximated via an analog to the first (and higher) order Onsager treatment of nonequilibrium thermodynamics, recognizing, however, that, due to the failure of information source microreversibility—e.g., 'eht' does not have the same probability as 'the' in English—there are no 'Onsager reciprocal relations' in high order approximations.
- For purely physical phenomena, free energy is often defined in terms of Boltzmann distribution-derived partition functions. In contrast, free energy analogs associated with the dynamics of cognitive phenomena can, and may well, be characterized by 'fat-tailed' distributions that do not have well-defined first or second moments (e.g., Derman et al. 2016; Wallace and Fricchione 2024).

Resource Rates

'Embodiment' assumes that cognitive systems of any nature are embedded in an environment that includes themselves. There must be at least three resource streams necessary for correct function:

- The rate at which elements of the system can communicate with each other, characterized by some channel capacity $\mathscr{C}$.
- The rate at which 'sensory information' is available from the embedding circumstance, according to a channel capacity $\mathscr{H}$.

- The rate at which 'material/materiel' resources can be provided, a rate $\mathscr{M}$. For an organism, this might be measured by the rate at which metabolic free energy is provided. Machine/institutional systems will require other measures.

These rates will, in a sense, cross-correlate or interact: one thus confronts a 3×3 (or larger) matrix $\mathbf{Z}$. An $n \times n$ matrix will have will have n scalar invariants r_i determined by the standard expression

$$p(\gamma) = \det[\mathbf{Z} - \gamma \mathbf{I}] =$$
$$(-1)^n \gamma^n + (-1)^{n-1} r_1 \gamma^{n-1} + ... - r_{n-1}\gamma + r_n \tag{1.1}$$

$\mathbf{I}$ is the appropriate identity matrix, 'det' is the determinant, and γ is a real-valued parameter. The first invariant is the matrix trace, and the last the determinant. We seek to construct a scalar resource rate measure from these invariants.

A direct, if simplistic, index might be $Z = \mathscr{C} \times \mathscr{H} \times \mathscr{M}$, but embodied AI worlds are not likely simple.

We assume—in first-order—that a single scalar index $Z(r_1, ..., r_n)$ can be constructed that sufficiently represents information and material resource rates. This approach is analogous to, but different from, the more usual principal component analysis based on a correlation matrix. Wallace (2021) outlines the complications associated with generalization to more than a single such index.

Rate Distortion Control Theory

A directly 'embodied' formalism emerges via a 'Rate Distortion Control Theory' based on the Rate Distortion Theorem. As described, the RDT addresses transmission under conditions of noise over an information channel. There will be, for that channel and a particular scalar measure of average distortion D between what is sent and what is received, a minimum necessary channel capacity $R(D)$.

The theorem asks, in effect, what is the 'best' channel for transmission of a message with the least possible average distortion. $R(D)$ can be defined for nonergodic information sources via a limit argument based on the ergodic decomposition of a nonergodic source into a 'sum' of ergodic sources. The RDT can be reconfigured in a control theory context, if we envision a system's topological information from the Data Rate Theorem as another form of noise, adding to the average distortion D. See Fig. 1.1.

The punctuation implied by the Data Rate Theorem (Nair et al. 2007) happens if there is a critical maximum average distortion beyond which control fails.

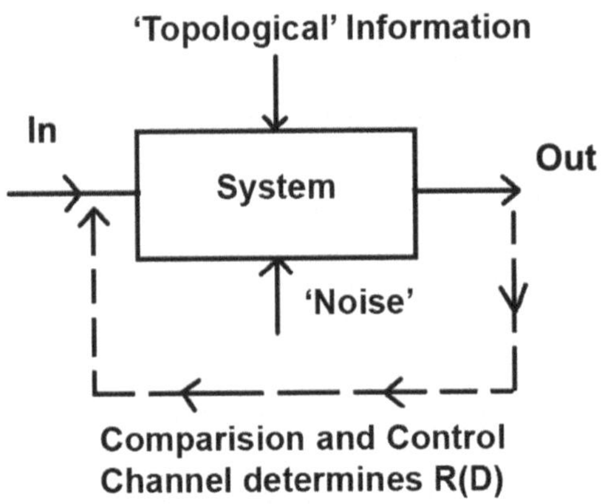

Fig. 1.1 The Rate Distortion Control Theory model. An inherently unstable system's 'topological information', according to the Data Rate Theorem (Nair et al. 2007), is added to the effects of 'noise' to determine the minimum channel capacity $R(D)$ needed to assure transmission with average distortion D. The punctuation implied by the DRT emerges here if there is a critical maximum average distortion beyond which control fails

1.4 The Context Model

The Basic Idea

We provide a basic approach, applying and abducting standard formalism from statistical mechanics (Landau and Lifshitz 2007)—given a proper definition of 'temperature'—for a cognitive system.

Consider the full ensemble of high probability developmental trajectories available to the system, written as Y_j, $j = 1, 2, \ldots$. Each trajectory is associated with a Rate Distortion Function-defined minimum necessary channel capacity R_j for a particular maximum average scalar distortion D_j. Then, assuming some basic underlying probability model having distribution $\rho_R(C_R, x)$, where C_R is a parameter set, we can define a pseudoprobability for a meaningful 'message' Y_j sent into the system of Fig. 1.1 as

$$dP = \frac{\rho_R(C_R, R/g(Z))dR}{\int_0^\infty \rho_R(C_R, R/g(Z))dR} = \frac{\rho_R(C_R, R_j/g(Z))dR}{g(Z)} \tag{1.2}$$

since, by definition, $\rho_R(C_R, x)$ is a probability density function that integrates to one on the interval $x = 0 \rightarrow \infty$.

$g(Z)$ is a classic 'temperature' analog that—here—must be further calculated from first principles, as it is taken as depending on the resource rate Z.

$g(Z)$ is, roughly speaking, the inverse of the 'rationality parameter' that controls the trade-off between utility and information processing cost in theories of bounded rationality (e.g., Genewein et al. 2015). For the record, however, an extensive search has failed to find any differential equation models directly deriving the Yerkes-Dodson Effect within the current Bounded Rationality literature. The approach we take below, by contrast, leads 'almost trivially' to equations for the Y-D Effect across a surprising number of modalities.

Here, for simplicity, we only treat the continuous case. For discrete or finite-state systems, the integral is a sum and the partition function denominator becomes some more general function $h(g(Z))$.

To reiterate, Y_j is some particular trajectory, while the (generalized) integral is over all possible 'high probability' paths.

We thus, at least implicitly, impose the Shannon-McMillan Source Coding Theorem (Cover and Thomas 2006): the set of possible system trajectories can be divided in two. These are, first, a very large set of measure zero—vanishingly low probability—that is not consistent with the underlying grammar and syntax of some basic information source, and a much smaller consistent set (Khinchin 1957).

Again, R_j is the RDT channel capacity keeping average distortion less than a given limit D_j for message Y_j. $g(Z)$ is a temperature analog depending on the scalar resource rate Z. $\rho(C, x)$, for physical systems, is usually taken as the Boltzmann distribution: $\rho(x) = \exp[-x]$. We suggest that, for cognitive phenomena—from the living and institutional to the machine and composite—it is necessary to move beyond analogs with physical systems. That is, it becomes necessary explore the influence of a variety of different probability distributions, including those with 'fat tails' (Derman et al. 2016; Taleb 2020), on the dynamics of cognition/regulation stability.

Again, standard methodology from statistical physics (Landau and Lifshitz 2007) identifies the denominator of Eq. (1.2) as a partition function, giving an iterated free energy-analog F as

$$\rho_F(C_F, F/g(Z)) \equiv \int_0^\infty \rho_R(C_R, R/g(Z))dR = g(Z)$$
$$g(Z) = \text{RootOf}\,(X - \rho_F(C_F, F/X)) \quad (1.3)$$

where 'RootOf' means setting the expression to zero and solving for X and ρ_F is an appropriate 'free energy' distribution that must be determined case-by-case for cognitive, as opposed to simple physical, phenomena. For physical phenomena, ρ_F can often be taken as the Boltzmann distribution. Again, C_X is simply an associated parameter set.

From chemical physics (Laidler 1987), we can define a cognition rate for the system as the reaction rate analog

$$L \equiv \frac{\int_{R_0}^\infty \rho_R(C_R, R/g(Z))dR}{\int_0^\infty \rho_R(C_R, R/g(Z))dR} = \frac{1}{g(Z)} \int_{R_0}^\infty \rho_R(C_R, R/g(Z))dR \quad (1.4)$$

R_0 is the minimum channel capacity needed to keep the average distortion below a critical value D_0 for the full system of Fig. 1.1.

The basic underlying probability model of Eq. (1.2)—via the partition function argument of Eq. (1.3)—determines system dynamics, but not system structure, and cannot be associated with a particular underlying network conformation, although these are related (Watts and Strogatz 1998; Barabasi and Albert 1999). More specifically, a set of markedly different networks may all be mapped on to any single given dynamic behavior pattern and, indeed, vice-versa, The same static network may display a spectrum of behaviors (Harush and Barzel 2017). We focus on dynamics rather than network topology, a shift from much current work.

Feynman (2000)—following (Bennett 1982)—identifies 'information' as a form of free energy, using Bennett's ideal machine that turns a message directly into work. We are concerned here with an iterated, rather than a direct, construction.

F, as a free energy, is subject to symmetry-breaking transitions as $g(Z)$ varies (Pettini 2007). These symmetry changes, however, are not as associated with physical phase transitions as represented by standard group algebras. Such symmetry changes represent transitions from playing one 'game' to playing another. For example, a cognitive system may engage in foraging behaviors that trigger a predatory attack by another system. Then the game changes from 'foraging' to 'escape'.

Thus 'cognitive phase changes' involve shifts between equivalence classes of high probability developmental/behavioral pathways that are represented as *groupoids*, a generalization of the group concept such that a product is not necessarily defined for every possible element pair, although multiple products with multiple identity elements are defined (Brown 1987; Weinstein 1996).

Dynamics emerge via a first-order Onsager approximation akin to that of nonequilibrium thermodynamics (de Groot and Mazur 1984) in the gradient of an entropy measure S constructed from the 'iterated free energy' F of Eq. (1.3) by means of a Legendre transform that should be familiar from the Black-Scholes approximation (Landau and Lifshitz 2007; Pettini 2007).

Recall from the (first-order) Onsager approximation of nonequilibrium thermodynamics that $\partial Z/\partial t \propto \partial S/\partial Z$. The full—again, standard, first-order—development is

$$
\begin{aligned}
S(Z) &\equiv -F(Z) + ZdF(Z)/dZ \\
\partial Z/\partial t &\approx dS/dZ = f(Z) \\
f(Z) &= Zd^2F/dZ^2 \\
F(Z) &= \int\int \frac{f(Z)}{Z} dZdZ + C_1 Z + C_2 \\
g(Z) &= \mathrm{RootOf}\left(X - \rho_F(C_F,\ F(Z)/X)\right)
\end{aligned}
\tag{1.5}
$$

An expression for $g(Z)$ follows from the second expression of Eq. (1.3). Note that the 'diffusion coefficient' has been implicitly set equal to 1 in the second part of Eq. (1.5).

Higher order approximations are in the same realm as higher order regression models, modulo dimensional consistency. Full treatment may involve generalization of the Legendre transform via formal power series methods (Jackson et al. 2017) .

There are several important points:

- Since 'RootOf' may have complex number solutions, the temperature analog $g(Z)$ enters the realm of the 'Fisher Zeros' characterizing phase transition in physical systems (e.g., Dolan et al. 2001; Fisher 1965; Ruelle 1964 Sect. 5). Such phase transitions lie at the base of the punctuated accession to consciousness in the Baars, and related, Global Workspace Models. Indeed, the 'RootOf' construction in Eq. (1.3) actually generalizes the Lambert W-function (e.g., Mezo and Keady 2015; Yi et al. 2010).
- To reiterate, information sources are not microreversible, that is, palindromes are highly improbable, for example the sequence 'ti' has far lower probability than 'it' in English. Thus, there are no 'Onsager Reciprocal Relations' in higher dimensional systems. The imposition of groupoid symmetries on cognitive phenomena appears to follow upon this directed homotopy.
- Further, there will always be a delay in the rate of provision of Z, so that, in Eq. (1.5), for example, $f(Z) = \beta - \alpha Z(t)$—an exponential model having $Z(t) = (\beta/\alpha)(1 - \exp[-\alpha t])$—where $Z \to \beta/\alpha$ at a rate determined by the 'friction' parameter α. Other dynamics are possible, such as the 'Arrhenius', $Z(t) = \beta \exp[-\alpha/t]$, with $f(Z) = (Z/\alpha)(\log(Z/\beta))^2$. However Z is constructed from the components $\mathcal{C}$, $\mathcal{H}$ and $\mathcal{M}$, its scalarization is a central matter, as higher dimensional constructs introduce serious mathematical complications (Wallace 2021).

Examples: Yerkes-Dodson Effects

We consider three examples for a 'contextual' cognitive process, based on the underlying distributions

$$
\begin{aligned}
\rho(x) &= \exp[-x] \\
\rho(x) &= 1/(1+x)^2 \\
\rho(x) &= \sqrt{\frac{1}{2\pi}} \frac{\exp[-1/(2x)]}{x^{3/2}}
\end{aligned}
\tag{1.6}
$$

The first is the Boltzmann Distribution having a 'hazard rate' of 1 and well-defined first and second moments. The second is fat-tailed, without finite first or second moments, but with the simple monotonically declining hazard rate $1/(1 + x)$. The third is the Levy Distribution, a classic fat-tailed distribution, again without finite first or second moments and having a quite complicated inverse-U hazard rate.

We will set $R_0 = 1$ in Eq. (1.4).

The relevant relations become, respectively,

For the Boltzmann Distribution:

$$\exp[-F/g] = g$$
$$g = \frac{-F}{W(n, -F)}$$
$$L = \exp[-1/g] \tag{1.7}$$

where $W(n, x)$ is the Lambert W-function satisfying the relation

$$W(n, x)\exp[W(n, x)] = x$$

It is real-valued only for $n = 0, -1$ and $x \geq -\exp[-1]$. We will take $n = 0$. For the $1/(1+x)^2$ distribution:

$$\frac{1}{(1+F/g)^2} = g$$
$$g = -F + \frac{1}{2} + \frac{\sqrt{-4F+1}}{2}$$
$$L = \frac{g}{1+g} \tag{1.8}$$

For the Levy Distribution:

$$\frac{e^{-\frac{g}{2F}}g}{\sqrt{2\pi}\,F\sqrt{\frac{F}{g}}} = g$$
$$g = -F\,W(0, -2F^2\pi)$$
$$L = \operatorname{erf}\left(\frac{1}{\sqrt{2/g}}\right) \tag{1.9}$$

where erf is the standard Error Function.

Again, we assume an 'exponential' model so that

$$dZ/dt = Zd^2F/dZ^2 = f(Z) = \beta - \alpha Z$$
$$F = \beta\,(Z\ln(Z) - Z) - \frac{\alpha\,Z^2}{2} + C_1 Z + C_2 \tag{1.10}$$

with $Z(t) \to \beta/\alpha$.

For Eq. (1.10), defining the 'free energy' as a function of Z, we set $\alpha = 1,\ C_1 = -1,\ C_2 = 3,\ Z = \beta$ and allow β to vary as the index of 'arousal' leading to the cognition rate analyses in Fig. 1.2. Figure 1.2a represents the cognition rate associated

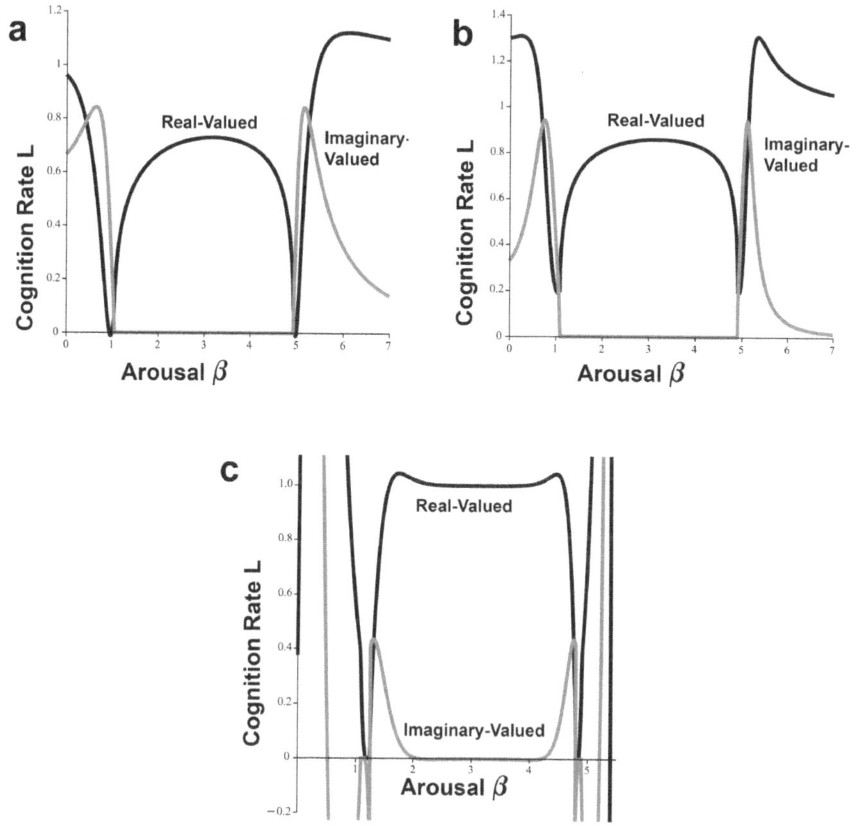

Fig. 1.2 Yerkes-Dodson inverse-U cognition rates for a 'contextual' system under three different underlying distributions. **a** Boltzmann Distribution, **b** $1/(1+x)^2$ distribution. **c** The Levy Distribution. The latter two do not have a finite mean, but for the 'context' model, all three display similar patterns of cognition rate versus arousal. All, however, are bookended by grossly unstable 'hallucination' and 'panic' modes having significant nonzero imaginary components

with the Boltzmann Distribution, Fig. 1.2b that of the $1/(1+x)^2$ distribution, and Fig. 1.2c for the Levy Distribution.

The darker traces are real-valued components, the lighter, imaginary-valued. Nonzero imaginary-valued cognition rates imply extreme instability, usually taken as 'hallucination' to the left of the inverse-U, and 'panic' to the right.

Here, for all distributions, with and without well-defined mean values, some version of the Yerkes-Dodson results of (Wallace and Fricchione 2024) appear. Even the 'highly pathological' Levy Distribution shows a near-zero imaginary component ($|x| \leq 0.003$) in the central range of arousal, suggesting some measure of relative stability.

Again, an extensive literature search failed to find any explicit differential equation models based on currently-popular 'bounded rationality' methods that explicitly express the Y-D Effect.

These outcomes follow from the 'context' implicit in Eq. (1.2).

1.5 The Salient Objects Model

Systems based in 'Western' cultural ideology/biology will focus on collections of salient events almost always at the exclusion of important contextual drivers. Wallace (2020) follows others in exploring the consequences of this systematic omission on important collective enterprise.

For a context-free 'singular events' system, the most elementary model is both absurdly simple and deadly subtle.

Imagine a very large parallel set of dual information sources having independent channel capacities $R_i(D_i)$ and define a probability distribution across the ensemble that replaces Eq. (1.2), along with a cognition rate that replaces Eq. (1.4), as

$$\begin{aligned} P_j &= \frac{R_j}{\sum_k R_k} = \frac{R_j}{R} \\ L &= \frac{1}{R} \sum_{R_j > R_0} R_j \end{aligned} \tag{1.11}$$

In the limit of very, very many possible channels this generalizes as

$$dP \sim \frac{R\rho_R(C, R/g)dR}{\int_0^\infty R\rho_R(C, R/g)dR} = \frac{R\rho_R(C, R/g)dR}{g^2 \langle R \rangle} \tag{1.12}$$

where $\rho_R(C, x)$ is an appropriate probability distribution across the system of concern, C is a defining parameter, and $\langle R \rangle$ is the mean value, if it indeed exists.

The fundamental observation is that the distribution ρ_R across the world of singular events must have a finite first moment for this approach to work, a matter that has obvious implications for the cognition rate. In particular, given a finite first moment, $g^2 \langle R \rangle$ becomes the partition function in a further iteration, where the relations of interest become

$$\begin{aligned} & \rho_F(F/g) = g^2 \langle R \rangle \\ L = \frac{\int_{R_0}^\infty R\rho_R(C, R/g)\, dR}{\int_0^\infty R\rho_R(C, R/g)\, dR} &= \frac{\int_{R_0}^\infty R\rho_R(C, R/g)\, dR}{g^2 \langle R \rangle} \end{aligned} \tag{1.13}$$

$\rho_F(F/g)$ is taken as representing an empirically-appropriate free energy distribution that may or may not be the same as ρ_R, and R_0 is the function-threshold channel capacity.

As above, the first expression can be solved for the 'temperature' parameter g in in terms of the free energy analog F, and this inserted into the expression for cognition rate L to define a function $L(\beta)$, analogous to Fig. 1.2, assuming Eq. (1.10).

Setting $\rho_R(C, x) = C \exp[-Cx]$ and $C^2 x \exp[-Cx]$—the 'simple' exponential and the second-order Erlang Distribution representing a two-stage decision process (Wallace 2025d)—we take $\rho_F(F/g) = \exp[-F/g]$ in both cases. The physically reasonable solutions are, respectively,

$$
\begin{aligned}
g &= -\frac{F}{2\,W\left(0, \frac{1}{2\sqrt{\frac{1}{F^2/C}}}\right)} \\
L &= \frac{1}{g}\exp[-CR_0/g]\,(CR_0 + g) \\
g &= -\frac{F}{2\,W\left(0, -\frac{\sqrt{2}}{2\sqrt{\frac{C}{F^2}}}\right)} \\
L &= \frac{\left(C^2R_0^2 + 2\,CgR_0 + 2\,g^2\right)\mathrm{e}^{-\frac{R_0C}{g}}}{2\,g^2}
\end{aligned}
\tag{1.14}
$$

Again, $W(n, x)$ is the Lambert W-function of order $n = 0$. Note that, in the first case, since the object of W is always ≥ 0, there can be no imaginary-valued components of g or L. The Erlang Distribution model, by contrast, can entertain complex-valued solutions.

Defining $F(\beta)$ as Eq. (1.10), and taking $C = R_0 = 1$, gives the cognition rate vs. arousal relation of Figs. 1.3a and b. For the exponential example, a surprisingly stable form of the Yerkes-Dodson pattern persists, again bracketed by 'hallucination' and 'panic', but without imaginary-valued components.

The second-order Erlang Distribution, representing a two-stage decision process, while displaying the central Y-D effect, is book-ended by complex-valued instability modes.

For extreme value 'fat-tailed' distributions, of course, cognition rate is indeterminate.

Here, for the 'simple' Exponential Distribution model, in contrast to the 'contextual system' analysis, for an appropriately 'stable' distribution $\rho_F(x)$—if the 'salient objects' system works at all—there may be no non-zero imaginary-valued components of cognition rate.

The second-order Erlang Distribution presents a more familiar case.

Much, then, depends on the probability model acting across the 'salient event/object' limit. If individual 'events/objects' can go haywire, producing extremes with significant probabilities—perhaps akin to autoimmune disease within a higher organism—overall systemic cognition collapses.

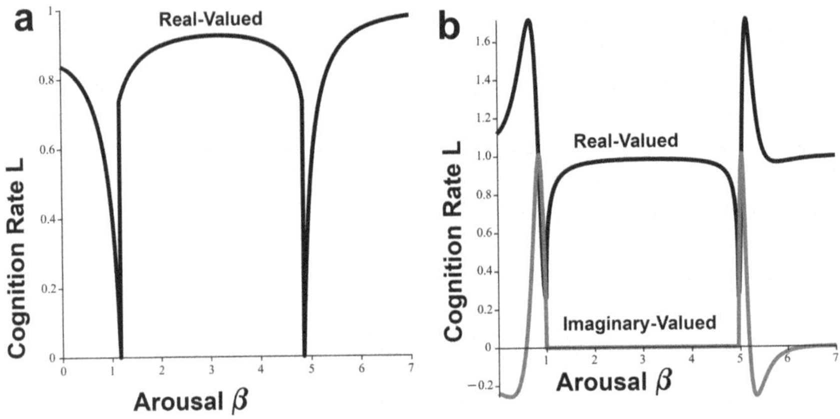

Fig. 1.3 Analog to Fig. 1.2 for the solvable 'salient objects' model based on the Exponential, second-order Erlang, and Boltzmann Distributions. **a** For the Exponential Distribution case, a surprisingly stable form of the Yerkes-Dodson pattern persists, again bracketed by 'hallucination' and 'panic', but without imaginary-valued components. **b** For the second-order Erlang Distribution, representing a two-stage decision process, complex-valued 'hallucination' and 'panic' modes recur. For extreme value 'fat-tailed' distributions, on the other hand, cognition rate becomes indeterminate

A standard and striking perceptual failure is, of course, the conclusion that some given distribution is 'well behaved' based on a limited set of observations. Failure to properly place 'salient events' within the proper embedding context, according to this modeling exercise, can lead to spectacular cognitive failure.

1.6 Fog and Friction Redux

Direct

The development leading to Eq. (1.5) can, more or less easily, be extended in a standard manner to the imposition of stochastic burden on the systems characterized by Eqs. (1.7) and (1.14). Then the relation $dZ/dt \sim f(Z)$ becomes the stochastic differential equation (Cyganowski et al. 2002)

$$dZ_t = f(Z_t)dt + \sigma h(Z_t)dB_t \tag{1.15}$$

where dB_t represents Brownian white noise and $h(Z)$ will be set to Z as 'ordinary' volatility. Then, using the Ito Chain Rule (see the chapter Mathematical Appendix for details), we can calculate the nonequilibrium steady state conditions $\langle dL_t \rangle = 0$, taking $f(Z) = \beta - \alpha Z$ set at the peak of Figs. 1.2a and 1.3, i.e., $\alpha = 1,\ \beta = 3,\ C_1 = -1,\ C_2 = 3,\ R_0 = 1$. Again, for the 'salient objects' system, $C = 1$.

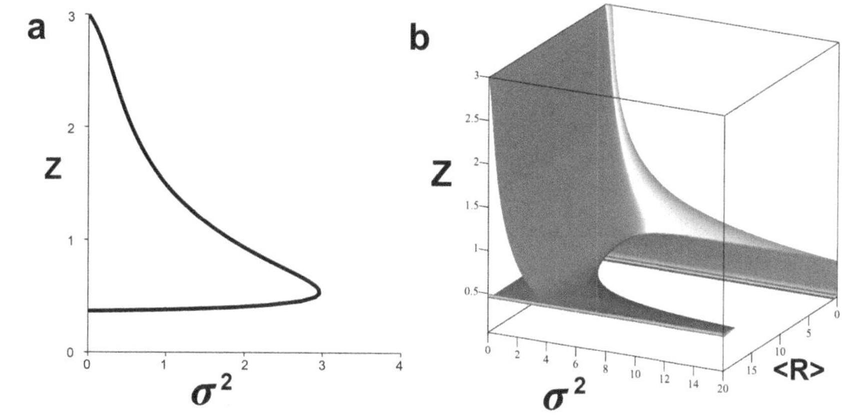

Fig. 1.4 **a** The principal mode for the 'contextual' exponential distribution model displays bifurcation and a degradation of $Z(\sigma^2)$ until a critical value for $\sigma^2 \approx 3$. **b** The salient objects system, while similarly degraded by rising 'noise', displays a complicated and counterintuitive dependence on $\langle R \rangle$

In Fig. 1.4a we calculate $Z(\sigma^2)$ and in Fig. 1.4b $Z(\sigma^2, \langle R \rangle)$ for the two (markedly) different expressions of the principal branches for $\langle dL_t \rangle = 0$ that follow from the formulas for cognition rate L in Eqs.(1.7) and (1.14).

The principal mode for the 'context' model displays bifurcation and a degradation of $Z(\sigma^2)$ until a critical value for $\sigma^2 \approx 3$. By contrast, the 'salient objects' system, while similarly degraded by rising 'noise', shows complicated and counterintuitive patterns depending on $\langle R \rangle$ as well as σ.

Again, Caveat Emptor.

Delayed

Previously, α in the expression $dZ/dt = \beta - \alpha Z(t)$ simply characterized a kind of 'friction'. This will likely be compounded with delay, here assumed to be a fixed quantity τ, so that $dZ/dt = \beta - \alpha Z(t - \tau)$. Imposing

$$Z(t) \equiv \frac{\beta}{\alpha}(1 - \exp[\mathscr{S}t]) \tag{1.16}$$

gives

$$\begin{aligned} \mathscr{S} &= \frac{W(n, -\alpha\tau)}{\tau} \\ dZ/dt &= \mathscr{S}(Z - \beta/\alpha) \\ dZ_t &= \mathscr{S}(Z_t - \beta/\alpha)\,dt + \sigma Z_t dB_t \end{aligned} \tag{1.17}$$

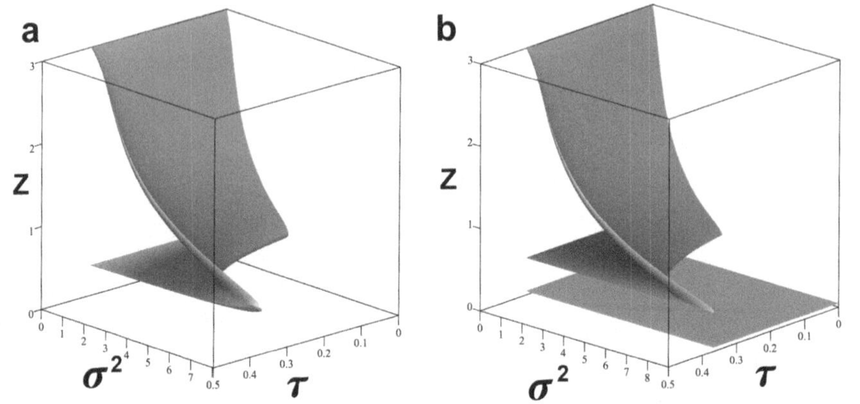

Fig. 1.5 We recalculate the nonequilibrium steady state conditions for the cognition rates $\langle dL_t \rangle = 0$ as in Fig. 1.4, setting $\langle R \rangle = 1$. **a** Context-driven system. **b** Salient object system. Since $\alpha = 1$ in both cases, the system fails as a phase transition at $\tau = \exp[-1] \approx 0.37$

$W(n, x)$ is again the n-th order Lambert W-function, assuming here $n = 0$, so that the synergism of 'friction' α with delay τ imposes a sharp phase transition when $\alpha\tau > \exp[-1]$.

We recalculate the nonequilibrium steady state conditions for cognition rates, $\langle dL_t \rangle = 0$, as in Fig. 1.4, fixing $\langle R \rangle = 1$, to give Figs. 1.5a and b. Note the phase transition to collapse for both at critical delay, in addition to that imposed by rising 'noise' σ^2.

Wallace (2022b) gives further details. Extension to distributed delay requires other methods.

Constrained

This argument can be further extended by recognizing that a 'system' is usually built from many interacting subcomponents that must operate under limited resources and time-to-act.

A relatively simple formal address can be made via 'ordinary' Lagrangian optimization.

We optimize cognition rates L_i of subcomponents under constraints of resource rates and time $Z = \sum_i Z_i$, $T = \sum_i T_i$ under the Lagrangian expressions

$$\mathcal{L} = \sum_i L_i + \lambda \left(Z - \sum_i Z_i \right) + \mu \left(T - \sum_i T_i \right)$$
$$\partial \mathcal{L} / \partial Z = \lambda, \; \partial \mathcal{L} / \partial T = \mu \tag{1.18}$$

so that, somewhat heuristically, $dZ/dT \sim \mu/\lambda$. λ and μ are *economic shadow prices* (Jin et al. 2008; Robinson 1993) representing the burden of additional environmental constraints.

Setting $dZ/dT = f(Z)$,

$$dZ/dT = \omega \equiv \frac{\mu}{\lambda} = f(Z)$$
$$Z(T) = f^{-1}(\omega) \quad (1.19)$$

Taking $f(Z) = \beta - \alpha Z$ or $\mathcal{S}(Z - \beta/\alpha)$, respectively,

$$Z = \frac{\beta - \omega}{\alpha}$$
$$Z = \frac{\beta}{\alpha} + \frac{\omega}{\mathcal{S}} \quad (1.20)$$

remembering that $\mathcal{S}$ from the delay calculation will be negative if real-valued and that $Z_i \leq Z$.

As ω rises, resources collapse.

Because $dZ/dT = \omega$, it is possible to reconsider the stochastic models above, condensing Eq. (1.15) as

$$dZ_t = \omega dt + \sigma Z_t dB_t \quad (1.21)$$

Taking the same conditions leading to Fig. 1.4 now leads to Fig. 1.6, where $\langle R \rangle = 1$.

The primary branches—the curved sections—display punctuated transitions at sufficient levels of 'fog' σ. The upper limits of the bifurcated surfaces cannot be reached under constraints on total Z.

In the Appendix we reconsider 'constraint' on resources delivered to a multi-component system from the perspective of a Shannon-Hartley reconfiguration of the model of Fig. 1.1, finding matters driven by a synergism between shadow price ratio, the 'time constant' of environmental events, and the 'noise' inherent to the overall control loop.

1.7 A Different Perspective

It is possible to continue this development in a different direction, invoking the Data Rate Theorem that characterizes the minimum control information needed to

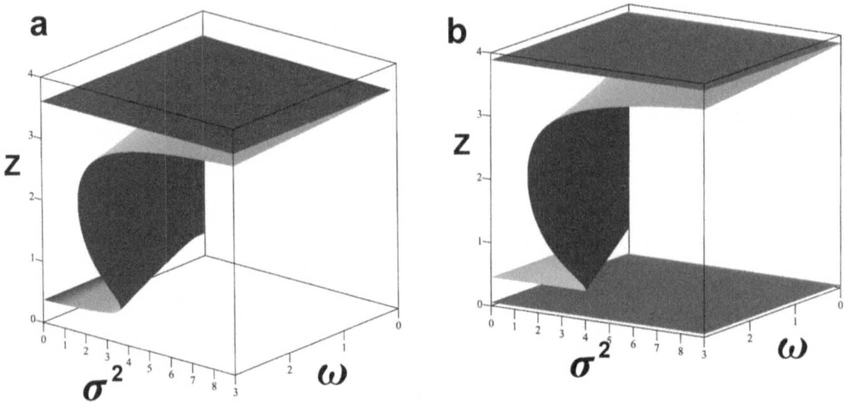

Fig. 1.6 Analog to Fig. 1.4, taking $\langle R \rangle = 1$, given resource and time constraints under shadow price burdens. **a** Context-Driven system. **b** Salient Objects system. Both principal branches show punctuated transitions under sufficient burden of 'fog' σ. The upper sections of the curved surfaces will not be reached under resource limitations

stabilize an inherently unstable control system. Nair et al. (2007) give the standard development, and Wallace (2025, Sects. 1.12 and 1.13) provides two simplified approaches.

The basic model is of a vehicle moving rapidly along a twisting roadway. The driver must provide control information—steering, braking, and changing gears—at an average distortion-limited rate $R(D)$ greater than the roadway can impose its sinuous 'topological information', say R_T, at the chosen average speed. Here, the limiting factor is the maximum allowable average distortion D. Recall that $R(D)$ is indeed a minimum information transmission rate defined by D.

We are particularly concerned with the 'salient events' model of Eqs. (1.12–1.13). The obvious conjecture is that the necessary condition for stabilization depends on a monotonically increasing function $Q(x)$ unbounded above on its domain such that

$$R > R_T \sim Q\left(\int_0^\infty R\rho_R(R/g)dR\right) = Q(g^2 \langle R \rangle) \tag{1.22}$$

If the distribution $\rho_R(R/g)$ is sufficiently 'fat-tailed', the mean $\langle R \rangle$ explodes and stabilization is not possible: there can never be a high enough rate of 'control information', a reverse fly-swatter or 'Babe Ruth' argument.

A generalization that addresses 'noise', based on the Black-Scholes model, is surprisingly straightforward, assuming a finite value for $\langle R \rangle$.

We can rewrite Eq. (1.22) as $R(Z) > R_T = \widehat{Q}(Z)R_0$ for some function $\widehat{Q}(Z)$. Using a Black-Scholes first-order approximation (Wallace 2020, Sect. 14.4), $R(Z) \approx \kappa_1 Z + \kappa_2$ and, again to first-order, $\widehat{Q}(Z) \approx \kappa_3 Z + \kappa_4$. Here, $\kappa_2 = 0$, and we can write a new 'temperature' relation for stability as

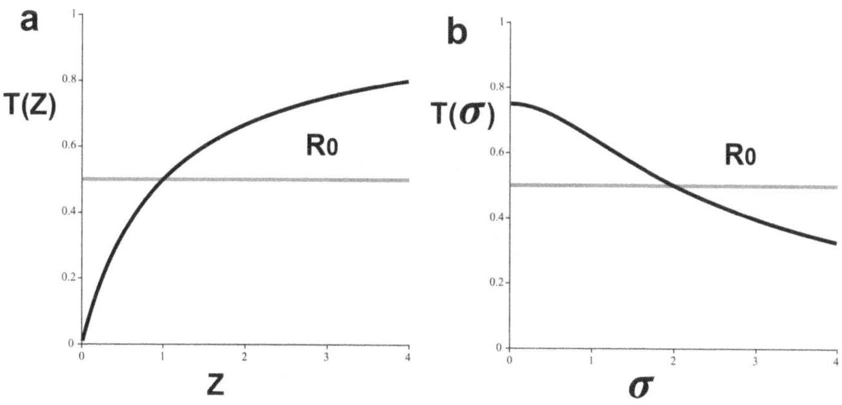

Fig. 1.7 **a** $T(Z)$, given a threshold R_0. **b** $T(\sigma)$ under the same threshold, fixing Z initially well above the critical level from (**a**)

$$T(Z) \equiv \frac{\kappa_1 Z}{\kappa_3 Z + \kappa_4} > R_0 \tag{1.23}$$

See Fig. 1.7a, where we set all the $\kappa_i = 1$ and $R_0 = 0.5$.

To impose noise we again assume a stochastic differential equation of the form

$$dZ_t = f(Z_t)dt + \sigma h(Z_t)dB_t \tag{1.24}$$

Note that, here, σ is taken as an added stochastic burden beyond that inherent to the standard definition of $R(D)$ that depends on the internal 'slop' of steering, braking, and gear-shifts.

Application of the Ito Chain Rule to the nonequilibrium steady state condition $\langle dT_t \rangle = 0$ gives

$$\sigma^2 = \frac{f(Z)}{h(Z)^2}\left(Z + \frac{\kappa_4}{\kappa_3}\right) \tag{1.25}$$

which can be back-solved for $Z(\sigma)$ to give an expression $T(\sigma)$.

To calculate Fig. 1.7b, we set $f(Z) = 3 - Z$, $h(Z) = Z$, $\kappa_4 = \kappa_3 = 1$, and solve for $T(\sigma)$, again with $R_0 = 0.5$. Enough 'noise' collapses the system below the stability threshold H_0.

Sufficient 'fat-tailed' unexpected events, and/or other 'noise', and the system cannot be stabilized.

For the exponential model under some discrete delay τ, $f(Z)$ in Eq. (1.25) is replaced by the second expression in Eq. (1.17), and matters are further compromised by the synergism between 'friction' α and τ that defines $\mathscr{S}$ in Eq. (1.17). In particular, $\mathscr{S}$ has non-zero imaginary-valued components if $\alpha\tau > \exp[-1]$. That is, even if

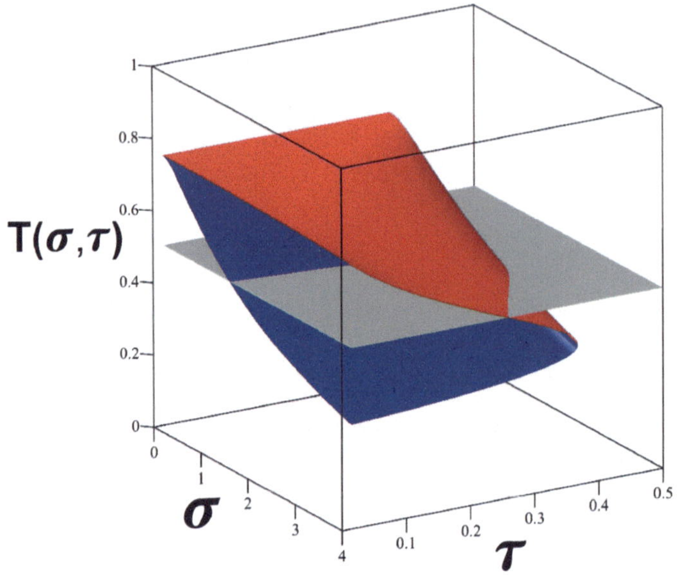

Fig. 1.8 Extension of the model of Fig. 1.7b under delay. There are two possible expressions for $\mathscr{S}$, corresponding to the $n = 0, -1$ forms of the Lambert W-function. These lead to the top and bottom sheaves/foliations of $T(\tau, \sigma)$. Since $\alpha = 1$, real-valued only solutions fail for $\tau > \exp[-1]$

extreme events are somehow suppressed, the interaction between delay and friction emerges as a central destabilizing factor.

Figure 1.8 displays the extension of Fig. 1.7b to this interaction, taking $n = 0, -1$ in the expression for $\mathscr{S}$, i.e., the two possible solution sets. Other parameters are the same. Again, since $\alpha = 1$, real-valued only solutions fail for $\tau > \exp[-1]$.

1.8 Intermediate Models

The fundamental difference in cognitive dynamics between the 'East Asian' and 'Western' models lies in the contrast between Eqs. (1.2) and (1.12). A more general 'full-spectrum' approach emerges if

$$dP \sim \frac{Q(R)\rho_R(C, R/g)dR}{\int_0^\infty Q(R)\rho_R(C, R/g)dR} \tag{1.26}$$

Then the 'Q' of this relation then determines overall dynamics, replacing Eqs. (1.3), (1.4) and (1.13) with the obvious generalizations.

For example, taking $\rho_F = \exp[-F/g]$ and $\rho_R = C\exp[-CR/g]$, but setting $Q(R) = R^n$, gives

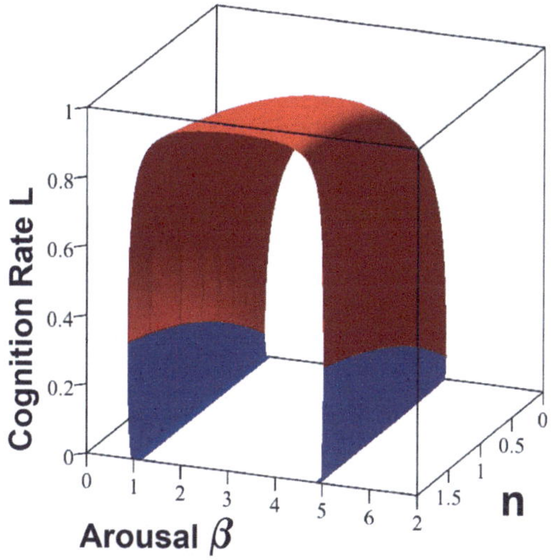

Fig. 1.9 Cognition rate $L(\beta, n)$ for the model of Eq. (1.27), setting $C = R_0 = 1$. For the Lambert W-function, $m = 0$ is the upper (red) sector, and $m = -1$ the lower (blue). The second-order Erlang Distribution model for ρ_R gives something similar

$$g = -\frac{F}{(n+1)\, W\left(m, -\frac{F C^{-\frac{n}{n+1}} \Gamma(n)^{\frac{1}{n+1}} n^{\frac{1}{n+1}}}{n+1}\right)}$$

$$L = 1 - \frac{g^{-\frac{n}{2}} C^{\frac{n}{2}} R_0^{\frac{n}{2}} \mathrm{e}^{-\frac{C R_0}{2g}} M_{\frac{n}{2}, \frac{n}{2}+\frac{1}{2}}\left(\frac{C R_0}{g}\right)}{\Gamma(2+n)} \tag{1.27}$$

Here, $W(m, x)$ is again the Lambert W-function of order m, and M is the Whittaker M-function of appropriate order.

Taking F as in Eq. (1.10), with $\alpha = 1$, $C_1 = -1$, $C_2 = 3$, Fig. 1.9 explores the cognition rate L as a function of arousal β and n for $Q(R) = R^n$. The upper (red) sector takes $m = 0$ for the Lambert W-function in the first expression of Eq. (1.27), and the lower (blue) sets $m = -1$. Here, the results are familiar, but again, for the general case, much will depend on the character of the distribution ρ_R.

A model based on the second-order Erlang Distribution for ρ_R, representing a 'two-stage exponential decision process', gives something similar to Fig. 1.9.

For the 'fat-tailed' Cauchy distribution

$$\rho_R(C, R) = \frac{2}{\pi C \left(1 + (R/C)^2\right)}$$

that does not have a finite first moment—again taking $\rho_F = \exp[-F/g]$ and $Q(R) = R^n$—the world changes. Then

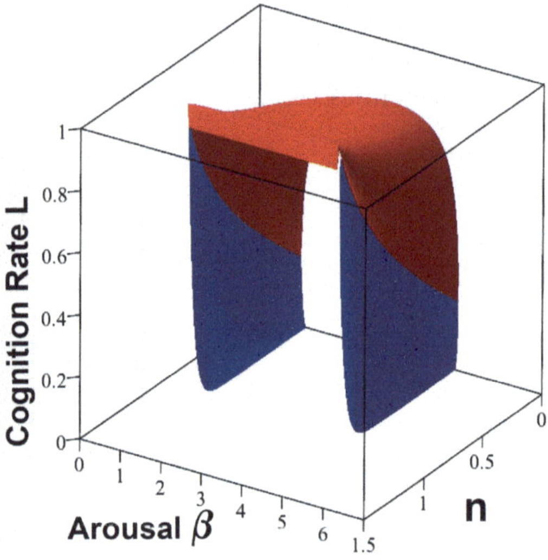

Fig. 1.10 Analog to Fig. 1.9 for the fat-tailed Cauchy distribution. Cognition collapses precipitously at $n = 1$ in the transition from 'East Asian' to 'Western' modes. Again the top and bottom (red and blue) sectors represent $m = 0, -1$ in the Lambert W-function

$$e^{-\frac{F}{g}} = \int_0^\infty R^n \rho_R(C, R/g) dR = g^{n+1} C^n \sec\left(\frac{\pi n}{2}\right)$$

$$g = -\frac{F}{(n+1)\, W\left(m, -\frac{F C^{\frac{n}{n+1}} \cos\left(\frac{\pi n}{2}\right)^{-\frac{1}{n+1}}}{n+1}\right)}$$

$$L = \frac{C^{1-n} g^{1-n} R_0^{n-1} \cos\left(\frac{\pi n}{2}\right) \Phi\left(-\frac{g^2 C^2}{R_0^2}, 1, \frac{1}{2} - \frac{n}{2}\right)}{\pi} \tag{1.28}$$

where Φ the Lerch Phi-function. An essential point is that $\sec(\pi n/2)$ in the first expression is singular at $n = 1$.

Setting $C = R_0 = 1$ and, as in Fig. 1.9, $m = 0, -1$ in the Lambert W-function, assuming F as in Fig. 1.9, gives Fig. 1.10, which is quite a different matter indeed: cognition rate collapses at $n = 1$ in the transition from 'East Asian' to 'Western' pattern.

For the distribution

$$\rho_R(C, x) = \frac{C}{(1 + Cx)^2}$$

which also does not have a finite first moment, the critical factor is found to be $n \csc(\pi n)$, where csc is the co-secant. The system again collapses at $n = 1$, following a close analog to Fig. 1.10.

A similar Levy Distribution calculation for ρ_R—again fat-tailed and without finite first moment—includes a multiplicative term in $\Gamma(1/2 - n)$. This, however, is singular at $n = 1/2$ instead of at $n = 1$, i.e., things can get rapidly nastier. Yet again, an analog to Fig. 1.10 emerges.

1.9 Is Any of This True?

A study by (Valmeekam et al. 2023) raised a singular red flag regarding the basic abilities of Large Language Models:

> Intrigued by the claims of emergent reasoning capabilities in LLMs trained on general web corpora... we set out to investigate... the effectiveness of LLMs in generating plans autonomously in commonsense planning tasks... We conduct a systematic study by generating a suite of instances on domains similar to the ones employed in the International Planning Competition... Our findings reveal that LLMs' ability to generate executable plans autonomously is rather limited, with the best model (GPT-4) having an average success rate of $\approx 12\%$ across the domains...

Shen et al. (2025) expanded matters significantly, studying LLM performance against task difficulty, recovering the Yerkes-Dodson Effect:

> ...[T]he impact of stress—a ubiquitous and critical factor in human cognitive processes—on LLM performance remains relatively unexplored. Understanding how LLMs respond to stress... holds profound theoretical signification for AI research, especially in exploring the robustness and adaptability of AI models...
>
> Our study demonstrates that LLMs exhibit performance patterns closely resembling those of humans under varying stress levels.

Using the 'Leaderboard 2 Benchmark' (Leaderboard 2024), Shen et al. developed prompts designed to induce varying levels of stress, derived from established psychological frameworks and calibrated on ratings from human participants. They applied these prompts to several LLMs to assess their responses across a range of tasks, including instruction-following, complex reasoning, and emotional intelligence. The findings indicate that LLMs, like humans, perform optimally under moderate stress, consistent with the Yerkes-Dodson law. Notably, their performance declines under both low and high-stress conditions.

Figure 1.11, adapted from Fig. 1 of (Shen et al. 2025), shows the performance of Llama-3-8B-Instruct on that benchmark, in comparison with the standard model of the Yerkes-Dodson Effect.

While Shen et al. (2025) suggest that large language models have 'captured and operationalized human-like stress-performance dynamics', this conclusion may, in fact, place the cart well before the horse. That is, the dynamics of cognition-under-stress across a broad range of phenomena appear to be constrained in like manner by the asymptotic limit theorems of information and control theories.

Previous work is consistent with this more global interpretation.

Sorensen et al. (2022), in a 'minimal model' neural network study of visual perception found

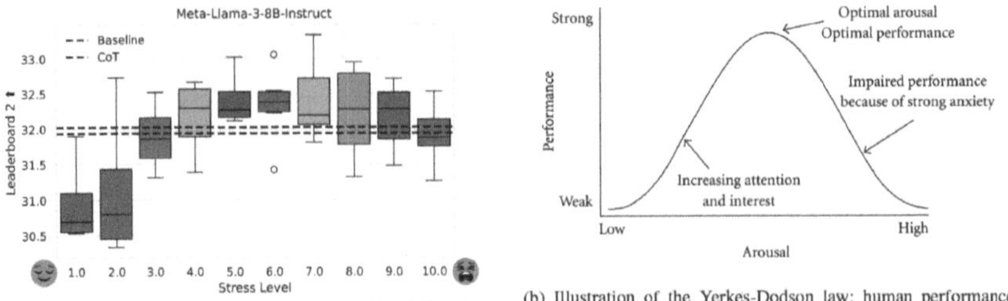

(a) Performance of `Llama-3-8B-Instruct` on Leaderboard 2 Benchmark (Leaderboard 2024) under different stress levels.

(b) Illustration of the Yerkes-Dodson law: human performance varies with stress levels, peaking at moderate stress and declining under low or high stress.

Fig. 1.11 Adapted from Fig. 1 of (Shen et al. 2025). The Yerkes-Dodson effect observed on a LLM

> ...[T]he Yerkes-Dodson effect... is not a by-product [of internal structural detail]... but rather reflects a global... mechanism... Our results... show that the Yerkes-Dodson effect can be replicated in our [neural network] model with a manipulation of perceptual difficulty.

Much earlier, a seminal study by French et al. (1999) explored the dynamics of a neural network model of the Yerkes-Dodson mechanism in explicit information theory terms,

> [Our] results suggest an interpretation of the Yerkes-Dodson law in terms of data compression. The capacity of a network to learn difficult tasks breaks down at high levels of arousal because the computation of network weights associated with the arousal and other inputs strains computational resources. Easy tasks require less computation and hence the network can perform well at higher levels of arousal.

Indeed, Wallace (2025b), following in these footsteps, explicitly invoked the Weber-Fechner, Stevens, Hick-Hyman and Pieron's sensory data compression psychophysics laws in different but roughly parallel derivations of the Yerkes-Dodson effect. See the Chapter Mathematical Appendix for an outline of the argument, based on stochastic extension of the models.

Moving beyond the current set of 'simple' LLMs, Shojace et al. (2025) challenge the more comprehensive Large Reasoning Models (LRMs) that have begun replacing them, using classic problems of increasing complexity. They find '...[F]rontier LRMs face a complete accuracy collapse beyond certain complexities... ultimately raising crucial questions about their true reasoning capabilities.'

Perhaps, however, not in quite the way Shojace et al. envision.

The upper panel of Fig. 1.12, taken from their Fig. 10, displays test accuracy results for three LRMs across four classic puzzles. The lower panel recalculates the example of Eqs. (1.7) and (1.10) from Fig. 1.2a, *changing the boundary conditions* of that example from $C_1 = -1,\ C_2 = 3$ to $C_1 = 2,\ C_2 = -7$, driving a critical transition at $\beta = 4.2105$.

Ultimately, it seems there is really no way around some form of signal transduction for 'difficult' cognitive problems across any milieu, context, or domain.

Some general insight can be gained by reconsidering the second expression of Eq. (1.7), defining the temperature analog g that follows from imposing the Boltzmann Distribution, i.e.,

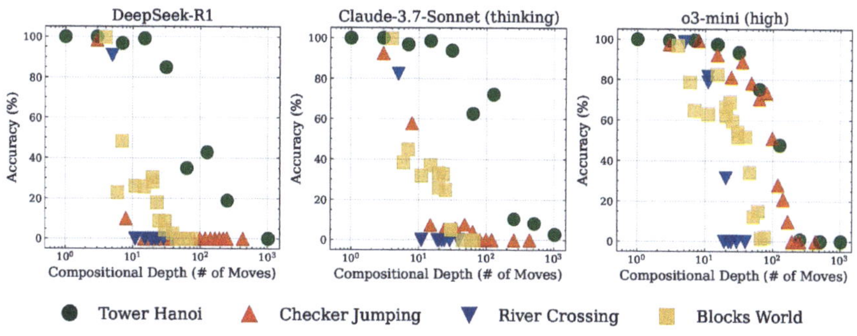

Accuracy versus compositional depth (number of moves required) for three LRMs (DeepSeek-R1, Claude-3.7-Sonnet with thinking, and o3-mini) across four puzzle environments.

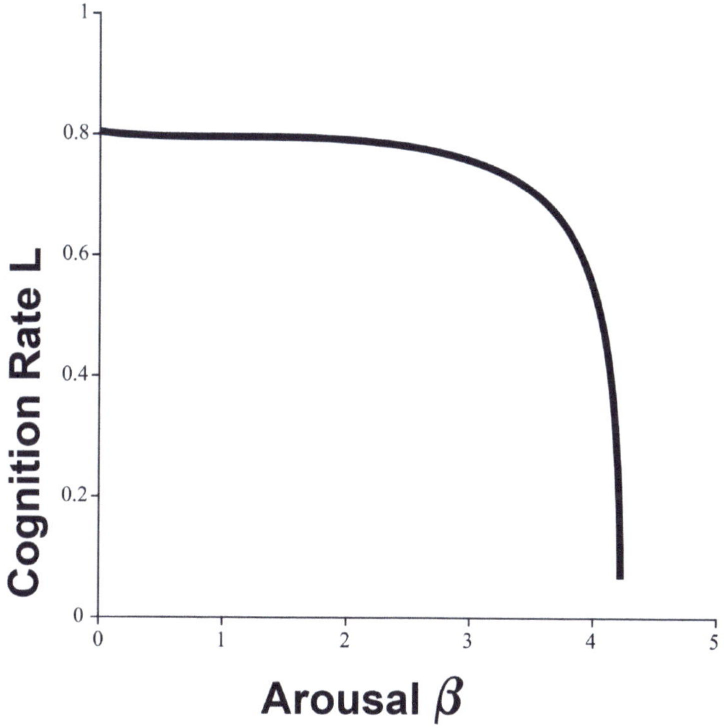

Fig. 1.12 The upper panel, from Fig. 10 of Shojace et al. (2025), displays observed accuracy for three LRMs across four classic puzzles. The lower recalculates the example of Eqs.(1.7) and (1.10) from Fig. 1.2a, changing the boundary conditions from $C_1 = -1$, $C_2 = 3$ to $C_1 = 2$, $C_2 = -7$. Here, the critical value is $\beta = 4.2105$. Some analogous form of signal transduction seems inevitable for 'difficult' cognitive problems across any mode

$$g = \frac{-F}{W(n, -F)}$$

The appearance of the Lambert W-function of orders $n = 0, -1$ places the restriction $-F \geq -\exp[-1]$ for real-valued g. The occurrence of a complex-valued 'temperature' introduces an analog to a 'Fisher-Zero phase transition' in physical systems (again, Dolan et al. 2001; Fisher 1965; Ruelle 1964 Sect. 5). Virtually all 'reasonable' probability distributions, leading to something like the last expression of Eq. (1.5), i.e., $\rho(F/g) = g$, drive the appearance of real-value range restrictions on F that will characterize a very broad range of 'cognitive' Fisher-Zero phase transitions. A few lines of clever mathematical reasoning should nail down the necessary and/or sufficient conditions for this limitation. Worked-out examples are provided in the Chapter Appendix.

1.10 Discussion

First, the examples studied here strongly imply that current (or forthcoming 'Third Wave') AI systems will not be any more successful at piercing Clausewitzian fog-and-friction and the burdens of attritional Zweikampf than have all the previous 'Revolutions in Military Affairs' making that claim (Wallace 2022b). After all, as Sun Tzu remarks, at the very least, the enemy always has a vote on the outcome.

Gray (1999), the Reagan Administration's go-to man on nuclear strategy, put some of it this way,

> The fog of war and frictions that harass and damage strategic performance do not comprise a static set of finite challenges which can be attrited by study, let alone by machines... The map of fog and friction is a living, dynamic one that reorganizes itself to frustrate the intrepid explorer.

A very particular implication of this work, however, is that—given the realities of Clausewitzian fog, friction, and deception—'Western' policies and polities that cultivate AI-enhanced whack-a-mole local tactical finesse at the arrogant and fulsome expense of contextual strategic insight and competence can fail spectacularly in the presence of extreme events that convert what had been perceived as a well-behaved probability model into something far different. The insistence that there is no forest, there are only trees, does not have survival value in a forest fire. The chill that the Wehrmacht felt outside Moscow in December 1941 was not only due to the cold, as a number of overextended and fading empires have similarly found in places like China, Vietnam, and Afghanistan, as did Pyrrhus in Italy, Varus in the Teutoburg Forest, and as the Global West and its client states do today in the Middle East.

No, the Leopard does not change its spots. The riverbanks of path-dependent historical trajectory channel the actions of AI—and other—cognitive systems built for critical purposes, obscuring their inevitable cultural blindspots. But even the strongest woven basket will not hold wine. Again, the enemy always gets a vote, and may focus that vote strategically, in full understanding of an opponent's persisting strengths and weaknesses. As has been said, 'Another such victory, and we are undone.'

Wallace (2021) views current attempts to use AI in organized conflict this way,

> [At the battle of Cynoscephale]...[a] Macedonian Phalanx, on rough ground and poorly protected by auxiliaries, was confronted by a far more flexible Roman legion. The Romans dispersed the auxiliary forces, flanked the formation and, with their short swords, slaughtered the rigidly-placed hoplites, hobbled by tiny shields and long lances...
>
> Into the world of Sun Tzu, Alexander, Scipio Africanus, Julius Caesar, Genghis Khan, Napoleon, Carl von Clausewitz, Robert E. Lee, U.S. Grant, Mikhail Tukhachevsky, Heinz Guderian, Erwin Rommel, Georgy Zhukov, Konstantin Rokossovsky, Mao Tse Tung, Vo Nguyen Giap, and John Boyd, marches the lock-step phalanx of AI researchers, their flanks protected by a straggling mob of camp-follower PR flacks, venture capitalists, careerists, and assorted fanboys.
>
> Beware the Romans: they don't fight fair.
>
> Nobody, really, fights fair...

Cognitive agents and agencies engaged in organized conflict are not simply constrained by the cultural riverbanks of their creation, but constitute a large part of those very riverbanks. As the increasingly mechanized, global wars of the 20th Century demonstrated, the resulting dynamics of variation and selection can become a 'language that speaks itself' (Wallace 2011; Chap. 5 of this book), independent of the survival of those embedded within it. Gray (2018) insists that the essence of strategy is prudence, finding the term 'nuclear strategy' to be an oxymoron. The blind embrace of culturally crafted and constrained AI into the management of critical human structures and systems may well become something similar.

A second exchange with the chatbot Perplexity AI in the Appendix regarding a military AI system currently offered by the US company Palantir is almost classic Galgen Humor black comedy. Even the chatbot gets it.

In sum, cultural blindness and determinism in organized conflict resound across all of history and will be entrained into the bones of all AI agents used in such enterprise. Singularly, to this author, the Tet Offensive of 1968, classified by US leadership as an egregious tactical and operational failure for the Viet Cong and the NVA, proved, ultimately, to be a spectacular strategic success.

Careerists—and incompetents—of course, will always claim that the light is just at the end of the tunnel, only a 'surge' away.

1.11 Appendix

On 'Embodiment': Summary Prepared by Perplexity AI

Evolutionary theory and contemporary AI research offer fundamentally different perspectives on the nature of intelligence, rooted in their respective disciplinary frameworks. This contrast revolves around embodiment, adaptation mechanisms, and definitions of problem-solving.

Evolutionary Perspective: Embodied Adaptation

Evolutionary biology frames intelligence as an emergent property of embodied systems interacting with dynamic environments. Key features include:

- Morphology as cognitive tool: Physical body shapes directly influence learning efficiency, as demonstrated by Stanford's 'unimals' study, where virtual creatures evolved better problem-solving abilities through iterative environmental challenges [1].
- Anticipatory adaptation: Natural selection enables systems to 'learn' from evolutionary history, allowing organisms to anticipate future pressures (e.g., bacteria evolving metabolic flexibility) [3,7].
- Decentralized cognition: Intelligence arises from distributed interactions between bodies, environments, and selective pressures rather than centralized processing [9].
  This aligns with Hubert Dreyfus' critique of early AI, which argued that intelligence cannot be reduced to abstract symbol manipulation but requires situated, embodied experience [2].

Contemporary AI Definitions: Task-Oriented Abstraction

Modern AI literature, dominated by computer science, emphasizes disembodied computational prowess:

- Functional mimicry: The ISO defines AI as 'machines that think, learn, and adapt,' focusing on replicating human-like outputs (e.g., language processing, image recognition) [6].
- Data-driven optimization: IBM characterizes AI as systems that 'simulate human learning' through pattern recognition in datasets, prioritizing predictive accuracy over physical interaction [8].
- Narrow task mastery: NASA's framework equates intelligence with autonomous task completion in 'varying circumstances,' such as decision support systems or robotic navigation [10].

Notably, only 9% of surveyed AI definitions in recent literature explicitly mention embodiment [4, 13], reflecting a prioritization of algorithmic efficiency over biological realism.

Key Contrasts

Aspect	Evolutionary View	AI Literature
Embodiment	Central to intelligence (body-environment coupling) [19]	Often incidental (e.g., GPT-4 lacks physical form) [12, 13]
Learning Mechanism	Selection pressures shape morphology/behavior [3,7]	Supervised/unsupervised training on static datasets [5, 8]
Scope	Broad (applies to bacteria, plants, animals) [7]	Anthropocentric (benchmarked against human tasks) [6,10]
Temporal Adaptation	Anticipatory (evolution "learns" from past) [3]	Reactive (models adjust to new data post-deployment) [11]

Bridging the Divide

Recent work in embodied AI (e.g., the Santa Fe Institute's predictive information maximization [9]) and evolutionary robotics seeks to reconcile these perspectives by evolving agents that self-organize through body-environment interactions. However, mainstream AI remains dominated by paradigms that abstract intelligence from its physical substrate-a tension Dreyfus foresaw in his critiques of computationalism [2, 13].

Citations:

1. https://neuroscience.stanford.edu/news/how-bodies-get-smarts-simulating-evolution-embodied-intelligence
2. https://davidcycleback.substack.com/p/the-1960s-clash-between-the-philosopher
3. https://www.southampton.ac.uk/news/2015/12/evolution-learning-theory-study.page
4. https://www.sciencedirect.com/science/article/pii/S0160289624000266
5. https://meng.uic.edu/news-stories/ai-artificial-intelligence-what-is-the-definition-of-ai-and-how-does-ai-work/
6. https://www.iso.org/artificial-intelligence/what-is-ai
7. https://pmc.ncbi.nlm.nih.gov/articles/PMC5413894/
8. https://www.ibm.com/think/topics/artificial-intelligence
9. https://www.santafe.edu/research/projects/theory-of-embodied-intelligence
10. https://www.nasa.gov/what-is-artificial-intelligence/
11. https://pmc.ncbi.nlm.nih.gov/articles/PMC8633405/
12. https://www.coursera.org/articles/what-is-artificial-intelligence
13. https://en.wikipedia.org/wiki/Artificial_intelligence
14. https://indiaai.gov.in/article/why-do-scientists-and-philosophers-agree-on-ai-hubert-dreyfus-s-views
15. https://pmc.ncbi.nlm.nih.gov/articles/PMC6963795/
16. https://www.jstor.org/stable/j.ctt2250vc6
17. https://www.4tu.nl/ethics/downloads/default/files/brey-2001-dreyfus.pdf
18. https://www.reddit.com/r/AskScienceDiscussion/comments/xhwx3z/does_evolution_always_lead_to_an_increase_in/
19. https://en.wikipedia.org/wiki/Embodied_cognition

20. https://www.youtube.com/watch?v=3UxX1mjkDRY
21. https://www.ncbi.nlm.nih.gov/books/NBK210002/
22. https://dspace.mit.edu/bitstream/handle/1721.1/6084/AIM-154.pdf
23. https://en.wikipedia.org/wiki/Evolution_of_human_intelligence
24. https://journals.sagepub.com/doi/10.1177/000271627441200104
25. https://www.scip.org/news/693878/The-Intelligence-of-Adaptation-How-Humans-Evolve-Their-Thinking.htm
26. https://www.britannica.com/technology/artificial-intelligence
27. https://www.gartner.com/en/topics/artificial-intelligence
28. https://www.sciencedirect.com/science/article/pii/S0268401221000761
29. https://guides.lib.uiowa.edu/c.php?g=132109&p=9917993
30. https://bsu.libguides.com/c.php?g=1395427&p=10320854
31. https://www.mckinsey.com/featured-insights/mckinsey-explainers/what-is-ai
32. https://cloud.google.com/learn/what-is-artificial-intelligence
33. https://www.csl.mtu.edu/winter98/cs320/AI/def.html
34. https://www.sciencedirect.com/science/article/abs/pii/S0893608013000117
35. https://cacm.acm.org/blogcacm/a-brief-history-of-embodied-artificial-intelligence-and-its-future-outlook/
36. https://a16z.com/podcast/bio-eats-world-evolving-embodied-intelligence/

'Constraint' Redux

We reconsider the multicomponent system of Eq. (1.18) in terms of a Shannon-Hartley model for Fig. 1.1. That is, Fig. 1.1 is now taken as involving the transmission of a message between a controlling agency and the system being controlled.

The corresponding Shannon-Hartley channel capacity C is then

$$C = B \ln \left(1 + \frac{ZT/\tau}{S^2} \right) \tag{1.29}$$

where B is the channel 'bandwidth', Z and T are the totals as in Eq. (1.18), τ is a characteristic time constant of environmental effects, and S^2 is the in-channel noise. The expression T/τ ensures dimensional consistency, since Z is the rate variate necessary for the S-H calculation.

By Eq. (1.18), under shadow price constraints, $dZ/dT = \omega$, $T = Z/\omega$, so that Eq. (1.29) becomes

$$C = B \ln \left(1 + \frac{Z^2}{\omega \tau S^2} \right) \tag{1.30}$$

again in the context of the base SDE of Eq. (1.21).

Using the Ito Chain Rule to calculate the nonequilibrium steady state condition $\langle dC_t \rangle = 0$ from Eq. (1.30), some algebra gives

$$\sigma^2 = \frac{2\omega \left(\omega \tau S^2 + Z^2 \right)}{Z \left(Z^2 - \omega \tau S^2 \right)} \tag{1.31}$$

Since $\sigma^2 \geq 0$, provided $\omega > 0$, a necessary condition for system stability is

$$Z > \sqrt{\omega \tau} S \tag{1.32}$$

driven by the synergism of 'control noise' S with the environmental characteristics ω and τ. Sufficient 'environmental burden' interacting with inherent control channel noise will obviate any possible value of resource rate Z. This outcome can, perhaps, be viewed as yet another form of the Data Rate Theorem.

It is interesting to note that, formally, if the shadow price ratio $\omega < 0$, the necessary condition becomes

$$Z < \sqrt{|\omega| \tau} S \tag{1.33}$$

strongly capping the stable performance range by structurally limiting the resource rate Z 'manageable' under load. Such scenarios may, in fact, be possible, i.e., 'fuse-blowing', particularly in the case of a 'rapid' environment having small characteristic time constant τ, perhaps consonant with John Boyd's 'OODA Loop' approach.

'Cognitive Distribution' Models

We follow the argument leading to the lower panel of Fig. 1.12 for the 'Cognitive Distribution'

$$\rho(x) = a x^m \exp[-b x^n]$$
$$\int_0^\infty \rho(x) dx = \frac{a\, b^{-\frac{m+1}{n}} \Gamma\left(\frac{m+1}{n}\right)}{n} = 1 \tag{1.34}$$

which permits explicit calculation of a.

The condition $\rho(F/g) = g$ arising from the Partition Function calculation can be solved for F as

$$F^n = -\frac{g^n m W\left(-\frac{bn\left(\frac{g b^{\frac{-m-1}{n}} \Gamma\left(\frac{m+1}{n}\right)}{n}\right)^{\frac{n}{m}}}{m}\right)}{bn} \tag{1.35}$$

where, yet again, $W(x)$ is the zero-order Lambert W-function that becomes imaginary-valued for $x < -\exp[-1]$. This condition, in turn, permits calculation of the critical value of g from the relation

$$-\frac{bn\left(\frac{g b^{\frac{-m-1}{n}} \Gamma\left(\frac{m+1}{n}\right)}{n}\right)^{\frac{n}{m}}}{m} = -e^{-1} \tag{1.36}$$

Substituting for g at criticality,

$$F^n = \frac{\left(\frac{nb^{\frac{1}{n}}}{\left(\frac{n}{m}\right)^{\frac{m}{n}} e^{\frac{m}{n}} \Gamma\left(\frac{m}{n}+\frac{1}{n}\right)}\right)^n m}{bn} \tag{1.37}$$

permitting calculation of the critical values of $F(Z)$.

The development for the 'Cauchy-like' distributions that generalize Pareto (Type II) or Lomax distributions, i.e.,

$$\rho(x) = \frac{C(n-1)}{(1+Cx)^n}$$
$$C > 0, \; n > 1 \tag{1.38}$$

is similar.

In particular, this distribution is heavy-tailed for $n \leq 3$, with finite mean only for $n > 2$ and finite variance for $n > 3$.

The hazard rate is, surprisingly, uniformly monotonic decreasing with increasing demand, giving this distribution a particular utility:

$$\frac{C(n-1)}{Cx+1} \tag{1.39}$$

The condition $\rho(F/g) = g$ leads to a relation analogous to Eq. (1.35):

$$F = \frac{g\left(C^{\frac{1}{n}}(n-1)^{\frac{1}{n}}\left(\frac{1}{g}\right)^{\frac{1}{n}} - 1\right)}{C} \tag{1.40}$$

The cognition rate is

$$L \equiv \frac{1}{g}\int_{R_0}^{\infty} \rho(R/g)dR = \frac{g^{n-1}}{(R_0C+g)^{n-1}}$$
$$g = \frac{CR_0L^{1/(n-1)}}{1-L^{1/(n-1)}} \tag{1.41}$$

Assuming F as in Eq. (1.10), we set $\alpha = 1, \; C_1 = -1, \; C_2 = 3$ and, for Eqs. (1.40) and (1.41), assign $C = R_0 = 1$. β again varies as the index of 'arousal'. Taking $n = 1 \rightarrow 3$, numerical solutions give Fig. 1.13a and b, classic inverse-U forms for both the temperature analog g and the cognition rate L, bracketed by imaginary-valued 'hallucination' and 'panic' modes.

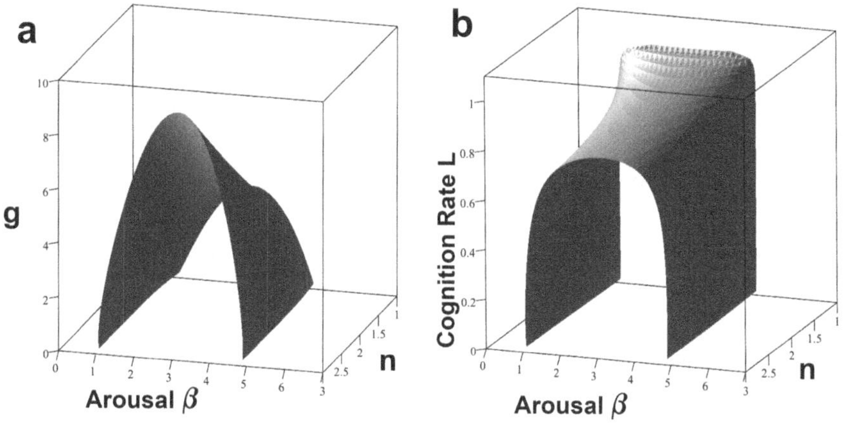

Fig. 1.13 Assuming F as in Eq. (1.10), $\alpha = 1$, $C_1 = -1$, $C_2 = 3$ and, in Eqs. (1.40), and (1.41), $C = R_0 = 1$. β again varies as the index of 'arousal', with $n = 1 \rightarrow 3$. Numerical solution gives **a** a classic inverse-U for the temperature analog g, and **b** the same for the cognition rate L, both bracketed by imaginary-valued 'hallucination' and 'panic' modes

Data Compression Models

It is possible to impose a different kind of order through the first expression of Eq. (1.10), via the stochastic differential equation

$$dZ_t = \left(Z_t d^2 F / dZ^2\right) dt + \sigma h(Z_t) dB_t \tag{1.42}$$

where, again, dB_t represents Brownian noise. Notice that we have not imposed $dZ/dt = f(Z)$, but need a further step to characterize the free energy measure F. This is done by requiring some signal compression 'law' $Q(Z)$ to be stabilized under the Ito Chain Rule. Calculation of $\langle dQ_t \rangle = 0$ gives the general relation

$$F(Z) = \int\int -\frac{\sigma^2 h(Z)^2 \left(\frac{d^2}{dZ^2} Q(Z)\right)}{2Z\left(\frac{d}{dZ} Q(Z)\right)} dZ dZ + C_1 Z + C_2 \tag{1.43}$$

The temperature analog $g(Z)$ and the corresponding cognition rate can then be calculated from Eqs.(1.3) and (1.4), according to the chosen probability distribution.

For Weber-Fechner data compression, $Q(Z) = \ln(1 + Z)$, taken under 'ordinary volatility' $h(Z) = Z$,

$$F = \frac{\sigma^2 \left(\frac{Z^2}{2} - \ln(1 + Z)(1 + Z) + 1 + Z\right)}{2} + C_1 Z + C_2 \tag{1.44}$$

We consider four examples.

1. The Boltzmann Distribution

$$\begin{aligned} \rho(x) &= \exp[-x] \\ g &= \frac{-F}{W(m, -F)} \\ L &= \exp[-R_0/g] \end{aligned} \tag{1.45}$$

2. The fat-tailed Lomax distribution

$$\begin{aligned} \rho(x) &= \frac{1}{(1+x)^2} \\ g &= -F + \frac{1}{2} + \frac{\sqrt{1-4F}}{2} \\ L &= \frac{g}{g+R_0} \end{aligned} \tag{1.46}$$

3. The second-order Erlang Distribution, representing a two-stage Boltzmann Distribution process

$$\begin{aligned} \rho(x) &= x\exp[-x] \\ g &= \frac{-F}{2W(-\sqrt{F}/2)} \\ L &= \frac{\exp[-R_0/g](R_0+g)}{g} \end{aligned} \tag{1.47}$$

4. The Cauchy Distribution, also fat-tailed

$$\begin{aligned} \rho(x) &= \frac{2}{\pi(1+x^2)} \\ g &= \frac{1+\sqrt{1-\pi^2F^2}}{\pi} \\ L &= 1 - \frac{1}{\pi}\arctan(R_0/g) \end{aligned} \tag{1.48}$$

Again, $W(m, x)$ is the Lambert W-function of orders $m = 0, -1$, real-valued only for $x > -\exp[-1]$.

Setting $R_0 = 1$, $C_1 = -1$, $C_2 = 3$ produces Figs. 1.14a–d, variously constrained by real-value limitations on F. There is, consequently, for all models, a maximum

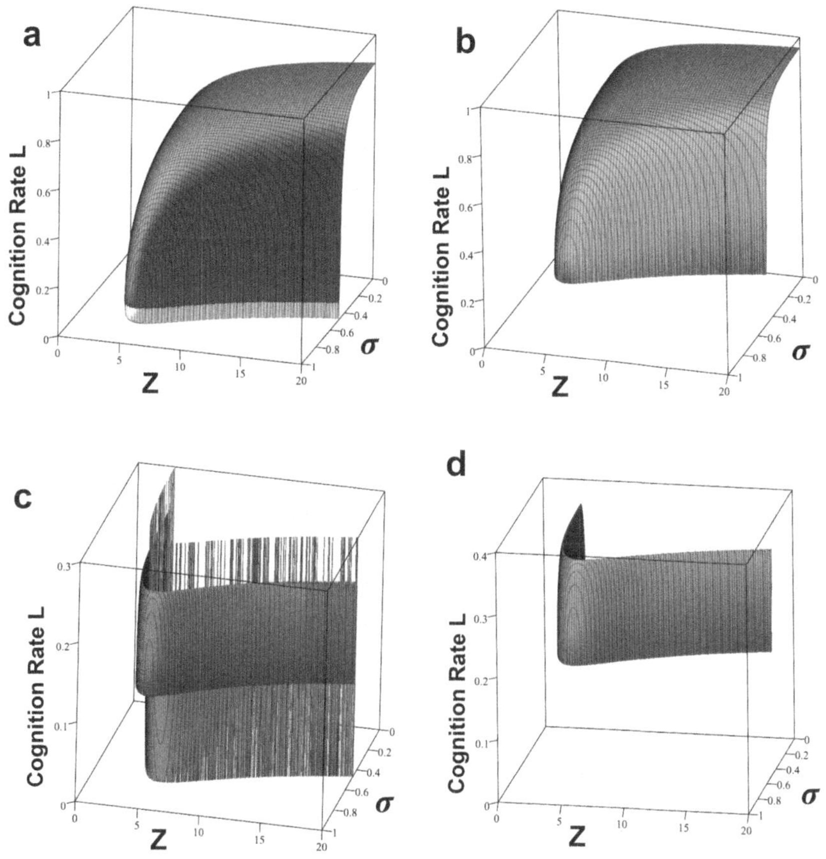

Fig. 1.14 For particularly 'well-behaved' underlying fundamental distributions, Weber-Fechner data compression in the presence of 'noise' can impose the Yerkes-Dodson Effect. Here $R_0 = 1$, $C_1 = -1$, $C_2 = 3$. **a** For the Boltzmann Distribution, the upper surface represents $m = 0$ and the lower $m = -1$ for the W-function. **b** Something roughly similar emerges for the $1/(1+x)^2$ distribution. **c** The second-order Erlang Distribution, representing a two-stage Boltzmann decision process, and the **d** Cauchy Distribution, present difficulties. In all cases, however, there is a phase transition defining a maximum tolerable level of 'noise'

tolerable stochastic burden σ, i.e., a phase transition. For the first two distributions, the Yerkes-Dodson Effect emerges, with 'easy' problems characterized by low σ. The last two distributions suggest the delicate dependencies of cognition on underlying distribution form.

There are, of course, many other possible 'sensory' data compression schemes, for example, again taking $h(Z) = Z$: Stevens' Law $Z^{1/n}$, $n > 1$, the Hick-Hyman Law, Z^{-1}, Pieron's Law, Z^{-b}, and the more technical compressions $\tanh(Z)$ and $\arctan(Z)$. These, respectively, lead to

$$F = \frac{\sigma^2(n-1)Z^2}{4n} + C_1 Z + C_2$$
$$F = \frac{1}{2}\sigma^2 Z^2 + C_1 Z + C_2$$
$$F = \frac{1}{4}\sigma^2(b+1)Z^2 + C_1 Z + C_2$$
$$F = \sigma^2\left(\frac{\mathrm{Li}_3(-e^{2Z})}{2} - \frac{Z^3}{6} - \frac{Z\,\mathrm{Li}_2(-e^{2Z})}{2}\right) + C_1 Z + C_2$$
$$F = \sigma^2\left(\frac{Z^2}{2} - Z\arctan(Z) + \frac{\ln(Z^2+1)}{2}\right) + C_1 Z + C_2 \tag{1.49}$$

where Li_n is the polylog function of appropriate order that also appears under logistic compression.

Detailed results, analogous to Fig. 1.14, are left as an exercise. The compression requirement, that $Q(Z) < Z$ for $Z > 0$, ensures that Eq. (1.43) generates phase transitions in σ when F drives the temperature analogs g and their associated cognition rates. Heuristically, this can be seen by inspection: For $Z > 1$, a compression scheme ensures $Q(Z) < Z$, so that $dQ/dZ < 1$ and $d^2Q/dZ^2 < 0$. Then $F = \sigma^2 u(Z) + C_1 Z + C_2$ with $u(Z) > 0$ for $Z > 0$.

Criticism by Perplexity AI of Palantir's AI Weapons Systems: Tactical Whack-A-Mole and Strategic Limitations

Overview of the Critique

A prominent criticism of Palantir's AI-powered weapons and military decision-support systems is that they offer only tactical, short-term solutions—akin to a game of 'whack-a-mole'—while failing to address the deeper, systemic issues of strategic incompetence within military or governmental organizations. Critics argue that these systems may create the illusion of technological dominance and control, but do not fundamentally improve the long-term strategic capacity or decision-making culture of their users [1, 2].

Key Arguments Behind the Criticism

- **Overemphasis on Tactical Fixes**
  Palantir's demos and marketing materials often showcase rapid, automated responses to battlefield threats—such as identifying enemy units and proposing immediate attack options. Critics point out that these are highly controlled scenarios that do not reflect the unpredictable, chaotic, and adaptive nature of real warfare [1, 2]
- **Illusion of Control**
  The platforms present a sanitized, 'perfect war' scenario where information is complete, the enemy is passive, and technology always works as intended. This can mask the reality that adversaries adapt, information is often incomplete or misleading, and technical systems are vulnerable to disruption or deception [2].

- **Strategic Incompetence Remains Untouched**
  By focusing on automating and accelerating tactical decisions, these AI systems do not address the root causes of strategic failure—such as poor leadership, flawed organizational culture, or lack of coherent long-term planning. The risk is that organizations become more efficient at making the same mistakes, rather than correcting them [1, 2].
- **Abstraction and Detachment**
  The automation and abstraction provided by Palantir's systems can further distance decision-makers from the realities and consequences of war, making it easier to overlook the political, ethical, and human dimensions of conflict [1, 2].
- **Dependency on Pristine Data and Infrastructure**
  The effectiveness of these AI systems relies on uninterrupted data flows, perfect integration, and the absence of adversarial interference. In real-world conditions, these assumptions rarely hold, and system failures can have catastrophic consequences [1, 2].
  Examples and Commentary
- **Military Demos Versus Reality**
  Palantir's own demos show operators using chatbots to order drone reconnaissance, generate attack plans, and jam enemy communications with minimal human input. In these scenarios, the 'human in the loop' often does little more than approve AI-generated actions, raising concerns about over-reliance on automated systems [1].
- **Expert Skepticism**
  Analysts and military experts have noted that such platforms offer 'the illusion of safety and control' but do not solve the 'various pernicious problems' of large language models (LLMs) and AI in military contexts, such as bias, brittleness, and susceptibility to adversarial manipulation [1, 2].
- **Selective Anaesthesia**
  Scholars have described these systems as providing a 'resilient numbness to the brute realities of warfare,' making violence and politics appear manageable and technical, rather than complex and deeply human [2].

Summary Table: Tactical vs. Strategic Impact

Aspect	Palantir AI Systems (Critics' View)	Long-Term Strategic Impact
Tactical Responsiveness	High—rapid, automated responses	Limited—may mask deeper problems
Strategic Competence	Not addressed by automation	Remains vulnerable to incompetence
Data & Environment Needs	Require pristine, uninterrupted data flows	Rarely met in real conflict
Human Decision-Making	Often reduced to approval of AI actions	Risk of detachment and over-trust
Adaptability	Assumes static, passive adversaries	Ignores enemy adaptation

Conclusion

The core of the criticism is that Palantir's AI weapons systems risk becoming sophisticated tools for fighting the last war, not preparing for the next. By focusing on tactical automation, they may entrench existing strategic weaknesses, rather than resolve them. This 'tactical whack-a-mole' approach, critics argue, cannot substitute for genuine improvements in strategic thinking, leadership, and organizational culture [1, 2].

1. https://www.business-humanrights.org/en/latest-news/palantir-claims-applying-generative-ai-to-warfare-is-ethical-without-addressing-problems-of-llms/
2. https://thebulletin.org/2023/08/war-is-messy-ai-cant-handle-it/

An Introduction to Stochastic Differential Equations

Einstein's famous 1905 analysis of Brownian motion for a large number of particles N goes something like this:

$$\begin{aligned} \partial\rho(x,t)/\partial t &= \mu\partial^2\rho(x,t)/\partial x^2 \\ \rho(x,t) &= \frac{N}{\sqrt{4\pi\mu t}}\exp[-\frac{x^2}{4\mu t}] \\ &< x^2 > \propto t \end{aligned} \tag{1.50}$$

where $\rho(t, x)$ is the probability of finding a particle at position x at time t, and μ is a 'diffusion coefficient'.

The last expression represents the average value of x^2 for a particle across the jittering system.

We can extend this insight to write, for a Brownian 'stochastic differential' dB_t that might be seen as perturbing some base function, the fundamental relation

$$\langle dB_t^2 \rangle = dt \tag{1.51}$$

Here, we follow Cyganowski et al. (2002, Sect.8.4), who provide a series of related programs for the computer algebra program MAPLE.

We take a base 'ordinary' differential equation $dX/dt = f(t, X(t))$ as perturbed by a Brownian stochastic variate dB_t that follows Eq. (1.51). The 'perturbed' solution X_t solves the stochastic differential equation (SDE)

$$dX_t = f(t, X_t)dt + g(t, X_t)dB_t \tag{1.52}$$

Suppose we are then given a function $Y(X_t, t)$ that depends explicitly on both t and X_t. Straightforward if tedious calculation—based on Eq. (1.52)—finds that $Y(t, X_t)$ solves the SDE

$$dY_t = L^0 Y_t dt + L^1 Y_t dB_t \tag{1.53}$$

where L^0 and L^1 are the operators

$$L^0Y = \partial Y/\partial t + f\partial Y/\partial x + \frac{1}{2}g^2\partial Y^2/\partial x^2$$
$$L^1Y = g\partial Y/\partial x \tag{1.54}$$

This is precisely the Ito Chain Rule.

To prove this, following Cyganowski et al. (2002, Sect. 8.4), one expands

$$\Delta Y_t = Y(t + \Delta t, X_t + \Delta X_t) - Y(t, X_t) \tag{1.55}$$

to second order, using $\Delta X_t = f\Delta t + g\Delta B_t + \cdots$. Further algebra generates a second-oder term in B_t

$$\frac{1}{2}g^2\partial^2 Y/\partial x^2(\Delta B_t)^2 \rightarrow$$
$$\frac{1}{2}g^2\partial^2 Y/\partial x^2 dt \tag{1.56}$$

by Eq. (1.51) that is then brought into the 'dt' part of the expression. This is one source of the characteristic strangeness of stochastic differential equations.

The Ito Chain Rule appears repeatedly in what follows.

We will be particularly concerned with 'nonequilibrium steady states' averaged across the stochastic jitter and represented as

$$\langle dY_t \rangle = L^0 Y = 0 \tag{1.57}$$

A MAPLE computer algebra program for the Ito Chain Rule can be found in Cyganowski et al. (2002, p. 238), shown here as

```
L0 := proc(X,a,b) local Lzero, U;
Lzero := diff(U(x,t),t)+a*diff(U(x,t),t,x)+
1/2 * b*b * diff(U(x,t), x, x);
eval(subs(U(x,t)=X,Lzero));
```

where X is the expression of Y in Eq. (1.54) in terms of x. a is $f(t, X_t)$ and b is $g(t, X_t)$ in Eq. (1.54), likewise expressed in terms of the base-variable x.

References

Atlan, H., Cohen, I.: Immune information, self-organization, and meaning. Int. Immunol. **10**, 711–717 (1998)

Bennett, C.H.: The thermodynamics of computation. Int. J. Theor. Phys. **21**, 905–940 (1982)

Borchert, H.: The very long game of defense AI adoption: introduction. In: Borchert, Schutz, Verbovsky (eds.) The Very Long Game: 25 Case Studies on the Global State of Defense AI. Springer, New York (2024)

Brown, R.: From groups to groupoids: a brief survey. Bull. Lond. Math. Soc. **19**, 113–134 (1987)

Champagnat, N., Ferriere, R., Meleard, S.: Unifying evolutionary dynamics: from individual stochastic process to macroscopic models. Theor. Popul. Biol. **69**, 297–321 (2006)

Clausewitz, K.: On War, ed. and trans. In: Howard, M., Paret, P. Princeton University Press, Princeton NJ (1976)

Cover, T., Thomas, J.: Elements of Information Theory, 2nd edn. Wiley, New York (2006)

Cyganowski, S., Kloeden, P., Ombach, J.: From Elementary Probability to Stochastic Differential Equations with MAPLE. Springer, New York (2002)

de Groot, S., Mazur, P.: Non-Equilibrium Thermodynamics. Dover, New York (1984)

Dembo, A., Zeitouni, O.: Large Deviations and Applications, 2nd edn. Springer, New York (1998)

Derman, E., Miller, M., Park, D.: The Volatility Smile. Wiley, New York (2016)

Dolan, B., Janke, W., Johnston, D., Stathakopoulos, M.: Thin Fisher zeros. J. Phys. A **34**, 6211–6223 (2001)

Dretske, F.: The explanatory role of information. Phil. Trans. R. Soc. A **349**, 59–70 (1994)

Dreyfus, H.: What Computers Can't Do: Of Artificial Reason. Harper and Row, New York (1972)

Dreyfus, H., Dreyfus, S.: Mind Over Machine: The Power of Human Intuition and Expertise in the Era of the Computer. The Free Press, New York (1986)

Farrell, H., Gopnik, A., Shalizi, C., Evans, J.: Large AI models are cultural and social technologies. Science **387**, 1153–1156 (2025)

Feynman, R.: Lectures on Computation. Westview Press, New York (2000)

Fisher, M.: Lectures in Theoretical Physics, vol. 7. University of Colorado Press, Boulder (1965)

Fjelland, R.: Why general artificial intelligence will not be realized. Hum. Soc. Sci. Commun. https://doi.org/10.1057/s41599–020-0494-4 (2020)

French, V., Anderson, E., Putman, G., Alvager, T.: The Yerkes-Dodson law simulated with an artificial neural network. Cogn. Syst. **5**, 136–147 (1999)

Genewein, T., Leibfried, F., Grau-Moya, J., Braun, D.: Bounded rationality, abstraction, and hierarchical decision-making: an information-theoretic optimality principle. Front. Robot. AI. https://doi.org/10.3389/frobt.2015.00027 (2015)

Gray, C.S.: Why strategy is difficult, pp. 7–12. Joint Force Quarterly, Summer (1999)

Gray, C.S.: Theory of Strategy. Oxford University Press, New York (2018)

Heine, S.: Self as cultural product: an examination of East Asian and North American selves. J. Pers. **69**, 881–906 (2001)

Jackson, D., Kempf, A., Morales, A.: A robust generalization of the Legendre transform for QFT. J. Phys. A **50**, 225201 (2017)

Jin, H., Hu, Z., Zhou, X.: A convex stochastic optimization problem arising from portfolio selection. Math. Financ. **18**, 171–183 (2008)

Jullien, F.: A Treatise on Efficacy: Between Western and Chinese Thinking. University of Hawaii Press, Honolulu (2004)

Khinchin, A.: The Mathematical Foundations of Information Theory. Dover, New York (1957)

Laidler, K.: Chemical Kinetics, 3rd edn. Harper and Row, New York (1987)

Landau, L., Lifshitz, E.: Statistical Physics, 3rd edn, Part 1. Elsevier, New York (2007)

Leaderboard, O.L.: Open-LLM performances are plateauing, let s make the leaderboard steep again (2024). https://huggingface.co/spaces/open-llm-leaderboard/blog. Accessed 16 Aug 2024

MacKenzie, D.: Da Vinci medal address: material political economy. Technol. Cult. **65**, 979–993 (2024). https://doi.org/10.1353/tech.2024.a933102

Markus, H., Kitayama, S.: Culture and the self: implications for cognition, emotion, and motivation. Psychol. Rev. **98**, 224–253 (1991)

Masuda, T., Nisbett, R.: Culture and change blindness. Cogn. Sci.: Multidiscip. J. **30**, 381–399 (2006)

Maturana, H., Varela, F.: Autopoiesis and Cognition: The Realization of the Living. Reidel, Boston (1980)

Mezo I., Keady, G.: Some physical applications of generalized Lambert functions (2015). arXiv:1505.01555v2 [math.CA] 22 Jun 2015

Nair, G., Fagnani, F., Zampieri, S., Evans, R.: Feedback control under data rate constraints: an overview. Proc. IEEE **95**, 108138 (2007)

Nisbett, R., Peng, K., Incheol, C., Norenzayan, A.: Culture and systems of thought: holistic versus analytic cognition. Psychol. Rev. **108**, 291–310 (2001)

Nisbett, R., Miyamoto, Y.: The influence of culture: holistic vs. analytic perception. Trends Cogn. Sci. **9**, 467–473 (2005)

Pettini, M.: Geometry and Topology in Hamiltonian Dynamics and Statistical Mechanics. Springer, New York (2007)

Richerson, P., Boyd, R.: Not by Genes Alone: How Culture Transformed Human Evolution. Chicago University Press, Chicago, IL (2006)

Robinson, S.: Shadow prices for measures of effectiveness II: general model. Oper. Res. **41**, 536–548 (1993)

Ruelle, D.: Cluster property of the correlation functions of classical gases. Rev. Mod. Phys. 580–584 (1964)

Shen, G., Zhao, D., Bao, A., He, X., Dong, Y., Zeng, Y.: StressPrompt: does stress impact large language models and human performance similarly? In: The Thirty-Ninth AAAI Conference on Artificial Intelligence (AAAI-25) (2025). Draft available at https://arxiv.org/html/2409.17167v2

Shojace, P., Mirzadeh, I., Alizadeh, K., Horton, M., Bengio, S., Farajtabar, M.: The illusion of thinking: understanding the strengths and limitations of reasoning models via the lens of problem complexity (2025). https://machinelearning.apple.com/research/illusion-of-thinking

Sorensen, L., Bohte, S., Slagter, H., Scholte, H.S.: Arousal state affects perceptual decision-making by modulating hierarchical sensory processing in a large-scale visual system model. PLOS Comput. Biol. (2022). https://doi.org/10.1371/journal.pcbi.1009976

Svensson, J.: Artificial intelligence is an oxymoron. AI and Society **38**, 363–372 (2023). https://doi.org/10.1007/s00146-021-01311-z

Valmeekam, K., Marquez, M., Sreedharan, S., Kambhampati, S.: On the planning abilities of large language models: a critical investigation. In: NIPS '23: Proceedings of the 37th International Conference on Neural Information Processing Systems, Article No.: 3320, pp. 75993–76005 (2023). https://dl.acm.org/doi/10.5555/3666122.3669442

Varnum, M., Grossman, I., Kitayama, S., Nisbett, R.: The origin of cultural differences in cognition: evidence for the social orientation hypothesis. Curr. Dir. Psychol. Sci. **19**, 9–13 (2010). https://doi.org/10.1177/0963721409359301

Wallace, R.: Culture and inattentional blindness: a global workspace perspective. J. Theor. Biol. **245**, 378–390 (2007)

Wallace, R.: A formal approach to evolution as self-referential language. Biosystems **106**, 36–44 (2011)

Wallace, R.: Cognitive Dynamics on Clausewitz Landscapes: The Control and Directed Evolution of Organized Conflict. Springer, New York (2020)

Wallace, R.: How AI founders on adversarial landscapes of fog and friction. J. Def. Model. Simul. (2021). https://doi.org/10.1177/1548512920962227

Wallace, R.: Consciousness, Cognition, and Crosstalk: The Evolutionary Exaptation of Nonergodic Groupoid Symmetry-Breaking. Springer, New York (2022a)

Wallace, R.: Deception and Delay in Organized Conflict: Essays on the Mathematical Theory of Maskirovka. Springer, New York (2022b)

Wallace, R.: Mathematical Essays on Embodied Cognition: Insights from Information and Control Theories. Springer, New York (2025a)
Wallace, R.: Sensory or intelligence data compression can drive the Yerkes Dodson Effect. Symmetry **17**, 235 (2025b). https://doi.org/10.3390/sym17020235
Wallace, R.: Hallucination and Panic in Autonomous Systems. Springer, New York (2025c)
Wallace, R.: Detailed Command Versus Mission Command: A Cancer-Stage Model of Institutional Decision-Making. Stats **8**, 27 (2025d). https://doi.org/10.3390/stats8020027
Wallace, R., Fricchione, G.: Stress-induced failure of embodied cognition: a general model. BioSystems **239**, 105193 (2024). https://www.sciencedirect.com/science/article/abs/pii/S0303264724000789
Weinstein, A.: Groupoids: unifying internal and external symmetry. Not. Am. Math. Assoc. **43**, 744–752 (1996)
Yi, S., Nelson, P.W., Ulsoy, A.G.: Time-Delay Systems: Analysis and Control Using the Lambert W Function. World Scientific, New Jersey (2010)
Zilberstein, S.: Using anytime algorithms in intelligent systems. AI Mag. **17**, 73–83 (1996)

Chapter 2
Stress, Uncertainty, and Cognitive Failure in Siloed Enterprise

2.1 Introduction

Financial Engineering and High Frequency Stock Trading have long been haunted by disruptions associated with 'fat tail' distributions that do not have well-defined first or second moments, that is, means or variances (Derman et al. 2016; Taleb 2020). Similarly, Wallace and Fricchione (2024) explored analogous disruptions that might afflict a relatively straightforward, if not entirely simple, model of biocognition, particularly under stress.

The essential point regards the conditions under which cognitive system dynamics become dominated by intrusive 'rare events' that, when all is said, cease to be sufficiently rare.

We begin with a brief recapitulation of the unsiloed 'single stream' cognition rate models of Wallace and Fricchione (2024), applied to both a well-behaved and a 'fat-tailed' distribution. Some fat-tailed distributions 'work' under the single-stream model, but both forms are bookended by intrusive dynamic instabilities.

By contrast, a fragmented, highly-parallel multi-stream 'siloed' configuration that may function under a 'regular' event stream, displays full cognitive collapse when dominated by 'extreme event' distributions without a well-defined mean.

From this perspective, a high-function, integrated enterprise would be most often characterized by the single cognitive stream: subsections and subgroups communicate well, working together to common purpose according to a shared, evolving, corporate doctrine. A pathological institution—or an ordinarily well-functioning institution overdriven by selection pressure—would be broken into parallel silos more concerned with self-preservation than collective enterprise, a grainy, crystalline structure in which subgroups have precipitated out under stress.

These are the two limits we will examine in some detail. Most real-world examples, however, will tend to be intermediate, and we outline an approach to the more complex dynamics of such systems.

© The Author(s), under exclusive license to Springer Nature Switzerland AG 2026

R. Wallace, *Bounded Rationality and Its Discontents*, Studies in Systems, Decision and Control 664, https://doi.org/10.1007/978-3-032-18423-8_2

The starting point for analysis involves characterization of cognition in terms of the asymptotic limit theorems of information and control theories. This is not entirely trivial and takes some attention.

2.2 Methodological Matters

The Asymptotic Limit Theorems

A singularly small set of asymptotic limit theorems from information and control theories constrains all possible cognitive phenomena, 'semantic' or not (Cover and Thomas 2006; Dretske 1994; Khinchin 1957; Nair et al. 2007). Building a new set of new statistical tools for the analysis of cognitive dynamics and dysfunctions, which is a central purpose of this analysis, depends on these theorems, just as the development of regression models, t-tests, and the like was based on the Central Limit Theorem, the Renewal Theorem, the Martingale Theorem, and so on.

The limit theorems of interest here are:

- The Shannon Coding Theorem that prescribes how to code a message in 'typical' sequences for maximum error-free transmission through a particular channel at a maximum 'channel capacity' (and the 'tuning theorem' variant in which the message formally 'transmits' the channel).
- The Source Coding Theorem that splits a set of sequences emitted by an information source into two subsets, a relatively small one that is 'meaningful'—the sequences are consonant with an underlying 'grammar' and 'syntax'—and a much larger 'meaningless' one 'of measure zero'. Only for stationary, ergodic systems is it possible to characterize 'information' and 'information source uncertainty' in terms of a Shannon entropy-analog. For nonergodic systems, each (infinite) signal sequence has it's own scalar source uncertainty that cannot be expressed in 'Shannon entropy' form (Khinchin 1957). It is here that methods must be abducted from statistical physics, introducing as well equivalence-classes of sequences forming a groupoid, leading to groupoid symmetry-breaking phase transitions characterizing 'cognitive' phase transitions, quite unlike those of 'conventional' physical systems.
- The Rate Distortion Theorem (RDT) and an associated Rate Distortion Control Theory that, in a sense, 'tunes' the channel capacity around a maximum-permissible scalar measure of average distortion (which finesses the Friston 'free energy minimization' approach).
- The Data Rate Theorem (DRT) stating that, for stability of an inherently unstable control system, 'control information' must be provided at a rate greater than the system produces it's own 'topological information'. A version of this is, somewhat surprisingly, 'easily' derived from a time and 'free energy rate' optimization based on the Hartley-Shannon Theorem.

After some further analysis:

- Cognition implies choice—of a small set of actions from a larger embedding set of what is possible to the system of interest. Such choice reduces uncertainty in a formal manner, implying the existence of an information source 'dual' to the cognitive process studied, at and across the varied scales and levels of organization of the living state (Atlan and Cohen 1998; Dretske 1994; Maturana and Varela 1980).
- Real-world, real-time cognition is always inherently unstable under the impact of the 'topological information' imposed by embedding environmental 'roadways'. The DRT implies that cognition must always be paired with regulation for system stability and long-term function under selection pressures. Indeed, it can perhaps be argued that the degree of pairing between cognition and regulation measures a system's 'self awareness', a central feature of many consciousness models.
- We impose an adiabatically, piecewise, stationary (APS) 'Born Oppenheimer' model in which the system of interest remains 'close enough' to stationary during the period(s) of interest so that the asymptotic limit theorems are effectively 'anytime algorithms' (e.g., Zilberstein 1996). Between 'pieces' for which this assumption is sufficiently true, we invoke (groupoid) symmetry-breaking and/or Fisher Zero (Dolan et al. 2001) phase transitions.
- The ergodic restriction can be lifted, at the cost of imposing methods abducted from statistical physics, with dynamics approximated via an analog to the first (and higher) order Onsager treatment of nonequilibrium thermodynamics, recognizing, however, that, due to the failure of information source microreversibility—e.g., 'eht' does not have the same probability as 'the' in English—there are no 'Onsager reciprocal relations' in high order approximations.
- For purely physical phenomena, free energy is often defined in terms of Boltzmann distribution-derived partition functions. In contrast, free energy analogs associated with the dynamics of cognitive phenomena can, and may well, be characterized by 'fat-tailed' distributions that do not have well-defined first or second moments (e.g., Derman et al. 2016; Wallace and Fricchione 2024).

Resource Rates

Embodiment recognizes that cognitive systems of any nature are embedded in an environment that includes themselves. There are at least three resource streams necessary for their function:

- The rate at which elements of the system can communicate with each other, instantiated by some channel capacity $\mathcal{C}$.
- The rate at which 'sensory information' is available from the embedding circumstance, say according to a channel capacity $\mathcal{H}$.
- The rate at which 'material/materiel' resources can be provided, a rate $\mathcal{M}$. For an organism, this might be measured by the rate at which metabolic free energy is provided. Machine/institutional systems will have other measures.

These rates will cross-correlate: one confronts a 3×3 (or larger) matrix $\mathbf{Z}$. An $n \times n$ matrix will have will have n scalar invariants r_i determined by the standard expression

$$p(\gamma) = \det[\mathbf{Z} - \gamma \mathbf{I}] =$$
$$(-1)^n \gamma^n + (-1)^{n-1} r_1 \gamma^{n-1} + \ldots - r_{n-1}\gamma + r_n \tag{2.1}$$

$\mathbf{I}$ is the appropriate identity matrix, 'det' is the determinant, and γ is a real-valued parameter. The first invariant is the matrix trace, and the last the determinant. We seek to construct a scalar resource rate measure from these invariants.

The most direct such index might be $Z = \mathcal{C} \times \mathcal{H} \times \mathcal{M}$, but our worlds are not likely so simple.

We postulate—in first-order—that a single scalar index $Z(r_1, \ldots, r_n)$ can be constructed that sufficiently represents information and material resource rates. This approach is analogous to, but different from, the more usual principal component analysis based on a correlation matrix. Wallace (2021) outlines the complications associated with generalization to more than a single such index.

Rate Distortion Control Theory

A directly 'embodied' formalism emerges via a 'Rate Distortion Control Theory' based on the Rate Distortion Theorem. As described, the RDT addresses transmission under conditions of noise over an information channel. There will be, for that channel and a particular scalar measure of average distortion D between what is sent and what is received, a minimum necessary channel capacity $R(D)$.

The theorem asks, in effect, what is the 'best' channel for transmission of a message with the least possible average distortion. $R(D)$ can be defined for nonergodic information sources via a limit argument based on the ergodic decomposition of a nonergodic source into a 'sum' of ergodic sources. The RDT can be reconfigured in a control theory context, if we envision a system's topological information from the Data Rate Theorem as another form of noise, adding to the average distortion D. See Fig. 2.1.

The punctuation implied by the Data Rate Theorem (Nair et al. 2007) happens if there is a critical maximum average distortion beyond which control fails.

2.3 The Single-Stream Model

We provide a basic approach, applying and abducting standard formalism from statistical mechanics (Landau and Lifshitz 2007)—given a proper definition of 'temperature'—for a cognitive system.

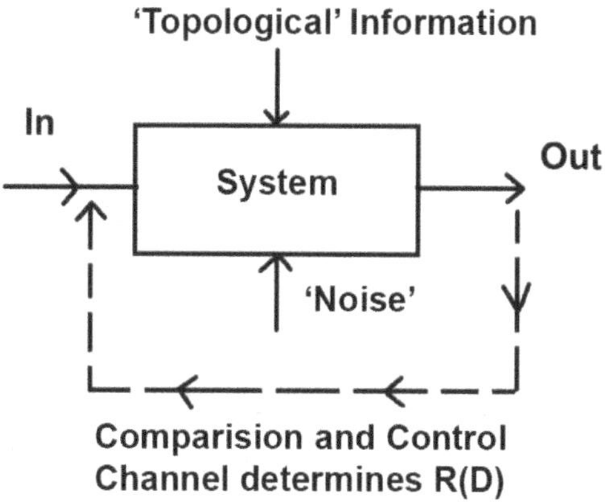

Fig. 2.1 The Rate Distortion Control Theory model. An inherently unstable system's 'topological information', according to the Data Rate Theorem (Nair et al. 2007), is added to the effects of 'noise' to determine the minimum channel capacity $R(D)$ needed to assure transmission with average distortion D. The punctuation implied by the DRT emerges here if there is a critical maximum average distortion beyond which control fails

Consider the full ensemble of high probability developmental trajectories available to the system, written as Y_j, $j = 1, 2, ...$ Each trajectory is associated with a Rate Distortion Function-defined minimum necessary channel capacity R_j for a particular maximum average scalar distortion D_j. Then, assuming some basic underlying probability model having distribution $\rho_R(C_R, x)$, where C_R is a parameter set, we can define a pseudoprobability for a meaningful 'message' Y_j sent into the system of Fig. 2.1 as

$$dP = \frac{\rho_R(C_R, R/g(Z))dR}{\int_0^\infty \rho_R(C_R, R/g(Z))dR} = \frac{\rho_R(C_R, R_j/g(Z))dR}{g(Z)} \tag{2.2}$$

since, by definition, $\rho_R(C_R, x)$ is a probability density function that integrates to one on the interval $x = 0 \to \infty$.

The 'temperature' $g(Z)$ in Eq. (2.2) deserves some comment. It is analogous to the inverse of the 'rationality parameter' of Bounded Rationality Theory (e.g., Genewein et al. 2015), but explicitly incorporates the resource rate index Z. Its functional form is not predetermined, but emerges from a first-principles calculation based on an iterated argument similar to Onsager's treatment of nonequilibrium thermodynamics. Recall that, according to (Feynman 2000)—following (Bennett 1982)—'information', and associated representations of channel capacity, are not 'entropies', but kinds of free energies from which an iterated actual entropy can then be defined. System dynamics arise from gradients of this 'real' entropy. Details will be given below for a first-order model. Higher order models can be imposed as empirically needed.

Here, for simplicity, we only treat the continuous case. For discrete or finite-state systems, the integral is a sum and the partition function denominator becomes some more general function $h(g(Z))$.

To reiterate, Y_j is some particular trajectory, while the (generalized) integral is over all possible 'high probability' paths.

We thus, at least implicitly, impose the Shannon-McMillan Source Coding Theorem (Cover and Thomas 2006): the set of possible system trajectories can be divided in two. These are, first, a very large set of measure zero—vanishingly low probability—that is not consistent with the underlying grammar and syntax of some basic information source, and a much smaller consistent set (Khinchin 1957).

Again, R_j is the RDT channel capacity keeping average distortion less than a given limit D_j for message Y_j. $g(Z)$ is the temperature analog from above, depending on the scalar resource rate Z. $\rho(C, x)$, for physical systems, is usually taken as the Boltzmann distribution: $\rho(x) = \exp[-x]$. We suggest that, for cognitive phenomena—from the living and institutional to the machine and composite—it is necessary to move beyond analogs with physical systems. That is, it becomes necessary explore the influence of a variety of different probability distributions, including those with 'fat tails' (Derman et al. 2016; Taleb 2020), on the dynamics of cognition/regulation stability.

Again, standard methodology from statistical physics (Landau and Lifshitz 2007) identifies the denominator of Eq. (2.2) as a partition function, giving an iterated free energy-analog F as

$$\rho_F(C_F, F/g(Z)) \equiv \int_0^\infty \rho_R(C_R, R/g(Z))dR = g(Z)$$

$$g(Z) = \text{RootOf}\,(X - \rho_F(C_F, F/X)) \tag{2.3}$$

where 'RootOf' means setting the expression to zero and solving for X and ρ_F is an appropriate 'free energy' distribution that must be determined case-by-case for cognitive, as opposed to simple physical, phenomena. For physical phenomena, ρ_F can often be taken as the Boltzmann distribution. Again, C_X is simply an associated parameter set.

From chemical physics (Laidler 1987), we can define a cognition rate for the system as the reaction rate analog

$$L \equiv \frac{\int_{R_0}^\infty \rho_R(C_R, R/g(Z))dR}{\int_0^\infty \rho_R(C_R, R/g(Z))dR} = \frac{1}{g(Z)} \int_{R_0}^\infty \rho_R(C_R, R/g(Z))dR \tag{2.4}$$

R_0 is the minimum channel capacity needed to keep the average distortion below a critical value D_0 for the full system of Fig. 2.1.

The basic underlying probability model of Eq. (2.2)—via the partition function argument of Eq. (2.3)—determines system dynamics, but not system structure, and cannot be associated with a particular underlying network conformation, although these are related (Barabasi and Albert 1999; Watts and Strogatz 1998). More specifically, a set of markedly different networks may all be mapped on to any single given dynamic behavior pattern and, indeed, vice-versa, The same static network may display a spectrum of behaviors (Harush and Barzel 2017). We focus on dynamics rather than network topology, a shift from much current work.

Feynman (2000)—following Bennett (1982)—identifies 'information' as a form of free energy, using Bennett's ideal machine that turns a message directly into work. We are concerned here with an iterated, rather than a direct, construction.

F, as a free energy, is subject to symmetry-breaking transitions as $g(Z)$ varies (Pettini 2007). These symmetry changes, however, are not as associated with physical phase transitions as represented by standard group algebras. Such symmetry changes represent transitions from playing one 'game' to playing another. For example, a cognitive system may engage in foraging behaviors that trigger a predatory attack by another system. Then the game changes from 'foraging' to 'escape'.

Thus 'cognitive phase changes' involve shifts between equivalence classes of high probability developmental/behavioral pathways that are represented as *groupoids*, a generalization of the group concept such that a product is not necessarily defined for every possible element pair, although multiple products with multiple identity elements are defined (Brown 1987; Weinstein 1996).

Dynamics emerge via a first-order Onsager approximation akin to that of nonequilibrium thermodynamics (de Groot and Mazur 1984) in the gradient of an entropy measure S constructed from the 'iterated free energy' F of Eq. (2.3) by means of a Legendre transform that should be familiar from the Black-Scholes approximation (Landau and Lifshitz 2007; Pettini 2007).

Recall from the (first-order) Onsager approximation of nonequilibrium thermodynamics that $\partial Z/\partial t \propto \partial S/\partial Z$. The full—again, standard, first-order—development is

$$
\begin{aligned}
S(Z) &\equiv -F(Z) + ZdF(Z)/dZ \\
\partial Z/\partial t &\approx dS/dZ = f(Z) \\
f(Z) &= Zd^2F/dZ^2 \\
F(Z) &= \int \int \frac{f(Z)}{Z} dZdZ + C_1 Z + C_2 \\
g(Z) &= \mathrm{RootOf}\,(X - \rho_F(C_F,\ F(Z)/X))
\end{aligned}
\tag{2.5}
$$

An expression for $g(Z)$ follows from the second expression of Eq. (2.3). Note that the 'diffusion coefficient' has been implicitly set equal to 1 in the second part of Eq. (2.5).

Again, higher order approximations are in the same realm as higher order regression models, modulo dimensional consistency. Full treatment may involve generalization of the Legendre transform via formal power series methods (Jackson et al. 2017).

There are several important points:

- Since 'RootOf' may have complex number solutions, the temperature analog $g(Z)$ enters the realm of the 'Fisher Zeros' characterizing phase transition in physical systems (e.g., Dolan et al. 2001; Fisher 1965; Ruelle 1964 Sect. 5). Such phase transitions lie at the base of the punctuated accession to consciousness in the Baars, and related, Global Workspace Models. Indeed, the 'RootOf' construction in Eq. (2.3) actually generalizes the Lambert W-function (e.g., Mezo and Keady 2015; Yi et al. 2010). It is the W-function and its generalizations that drive the inverse-U 'Yerkes-Dodson' stress responses.
- To reiterate, information sources are not microreversible, that is, palindromes are highly improbable, for example the sequence 'ti' has far lower probability than 'it' in English. Thus, there are no 'Onsager Reciprocal Relations' in higher dimensional systems. The imposition of groupoid symmetries on cognitive phenomena appears to follow upon this directed homotopy.
- Further, there will always be a delay in the rate of provision of Z, so that, in Eq. (2.5), for example, $f(Z) = \beta - \alpha Z(t)$—an exponential model having $Z(t) = (\beta/\alpha)(1 - \exp[-\alpha t])$—where $Z \to \beta/\alpha$ at a rate determined by α. Other dynamics are possible, such as the 'Arrhenius', $Z(t) = \beta \exp[-\alpha/t]$, with $f(Z) = (Z/\alpha)(\log(Z/\beta))^2$. However Z is constructed from the components $\mathcal{C}$, $\mathcal{H}$ and $\mathcal{M}$, so here, it is the scalar resource rate Z itself that concerns us.

We consider three examples for a single-stream cognitive process, based on the underlying distributions

$$\rho(x) = \exp[-x]$$
$$\rho(x) = 1/(1+x)^2$$
$$\rho(x) = \sqrt{\frac{1}{2\pi}}\frac{\exp[-1/(2x)]}{x^{3/2}} \tag{2.6}$$

The first is the Boltzmann Distribution having a 'hazard rate' of 1 and well-defined first and second moments. The second is fat-tailed, without finite first or second moments, but with the simple monotonically declining hazard rate $1/(1+x)$. The third is the Levy Distribution, a classic fat-tailed distribution, again without finite first or second moments and having a quite complicated inverse-U hazard rate.

We will set $R_0 = 1$ in Eq. (2.4).

The relevant relations become, respectively,

For the Boltzmann Distribution:

$$\exp[-F/g] = g$$
$$g = \frac{-F}{W(n, -F)}$$
$$L = \exp[-1/g] \tag{2.7}$$

where $W(n, x)$ is the Lambert W-function satisfying the relation

$$W(n, x)\exp[W(n, x)] = x$$

It is real-valued only for $n = 0, -1$ and $x \geq -\exp[-1]$. We will take $n = 0$. For the $1/(1+x)^2$ distribution:

$$\frac{1}{(1+F/g)^2} = g$$
$$g = -F + \frac{1}{2} + \frac{\sqrt{-4F+1}}{2}$$
$$L = \frac{g}{1+g} \tag{2.8}$$

For the Levy Distribution:

$$\frac{e^{-\frac{g}{2F}} g}{\sqrt{2\pi}\, F \sqrt{\frac{F}{g}}} = g$$
$$g = -F W(0, -2F^2\pi)$$
$$L = \mathrm{erf}\left(\frac{1}{\sqrt{2/g}}\right) \tag{2.9}$$

where erf is the standard Error Function.

Again, we assume an 'exponential' model so that

$$dZ/dt = Zd^2F/dZ^2 = f(Z) = \beta - \alpha Z$$
$$F = \beta\,(Z\ln(Z) - Z) - \frac{\alpha Z^2}{2} + C_1 Z + C_2 \tag{2.10}$$

with $Z(t) \to \beta/\alpha$.

For Eq. (2.10), defining the 'free energy' as a function of Z, we set $\alpha = 1$, $C_1 = -1$, $C_2 = 3$, $Z = \beta$ and allow β to vary as the index of 'arousal' leading to the cognition rate analyses in Fig. 2.2. Figure 2.2a represents the cognition rate associated with the Boltzmann Distribution, Fig. 2.2b that of the $1/(1+x)^2$ distribution, and Fig. 2.2c for the Levy Distribution.

The darker traces are real-valued components, the lighter, imaginary-valued. Nonzero imaginary-valued cognition rates imply extreme instability, usually taken as 'hallucination' to the left of the inverse-U, and 'panic' to the right. Here, for all

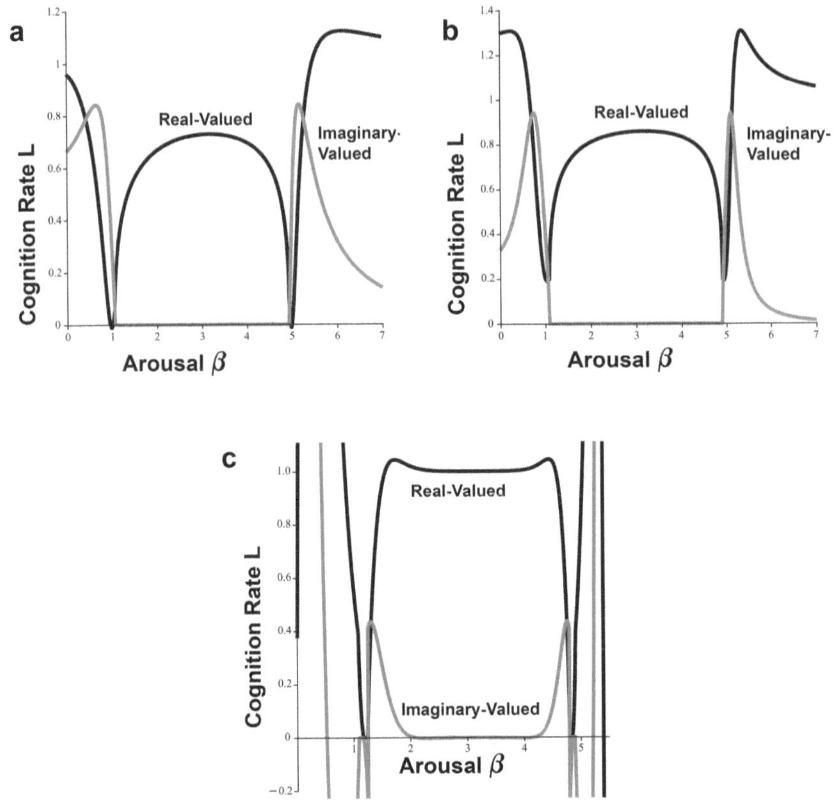

Fig. 2.2 Yerkes-Dodson inverse-U cognition rates for a single-stream system under three different underlying distributions. **a** Boltzmann Distribution, **b** $1/(1+x)^2$ distribution. **c** The Levy Distribution. The latter two do not have a finite mean value, but for the single-stream model, all three systems display similar patterns of cognition rate versus arousal. All three, however, are book-ended by grossly unstable 'hallucination' and 'panic' modes having significant nonzero imaginary components

distributions, with and without well-defined mean values, some version of the Yerkes-Dodson results of (Wallace and Fricchione 2024) appear. Even the 'highly pathological' Levy Distribution shows a near-zero imaginary component ($|x| \leq 0.003$) in the central range of arousal, suggesting some measure of relative stability.

These outcomes follow from the single-stream implicit in Eq. (2.2).

2.4 The Siloed Multi-stream Model

Real-world cognition is often a multi-stream conglomerate, a composite of Darwinian or Lamarckian selection processes within an institution, of competing demands within an organism, and so on. Flight-or-fight, available resources versus the demands of environmental affordance or threat, and other such, come to mind.

For a siloed multi-stream system, the most elementary model is both absurdly simple and deadly subtle.

Imagine a very large parallel set of dual information sources having independent channel capacities $R_i(D_i)$ and define a probability distribution across the ensemble that replaces Eq. (2.2), along with a cognition rate that replaces Eq. (2.4), as

$$
\begin{aligned}
P_j &= \frac{R_j}{\sum_k R_k} = \frac{R_j}{R} \\
L &= \frac{1}{R} \sum_{R_j > R_0} R_j
\end{aligned}
\tag{2.11}
$$

In the limit of very, very many possible channels this generalizes as

$$
dP \sim \frac{R\rho_R(C, R/g)dR}{\int_0^\infty R\rho_R(C, R/g)dR} = \frac{R\rho_R(C, R/g)dR}{g^2 \langle R \rangle} \tag{2.12}
$$

where $\rho_R(C, x)$ is an appropriate probability distribution across the institution, C is a defining parameter, and $\langle R \rangle$ is the mean value, if it indeed exists.

The fundamental observation is that the distribution ρ_R across the siloed enterprise must have a finite first moment for this approach to work, a matter that has obvious implications for the cognition rate. In particular, given a finite first moment, $g^2 \langle R \rangle$ becomes the partition function in a further iteration, where the relations of interest become

$$
\begin{aligned}
&\rho_F(F/g) = g^2 \langle R \rangle \\
L = \frac{\int_{R_0}^\infty R\rho_R(C, R/g)\,dR}{\int_0^\infty R\rho_R(C, R/g)\,dR} &= \frac{\int_{R_0}^\infty R\rho_R(C, R/g)\,dR}{g^2 \langle R \rangle}
\end{aligned}
\tag{2.13}
$$

$\rho_F(F/g)$ is taken as representing an empirically-appropriate free energy distribution that may or may not be the same as ρ_R, and R_0 is the function threshold channel capacity.

As above, the first expression can be solved for the 'temperature' parameter g in in terms of the free energy analog F, and this inserted into the expression for cognition rate L to define a function $L(\beta)$, analogous to Fig. 2.2, assuming Eq. (2.9).

Respectively setting

$$
\rho_R = C \exp[-Cx], \quad \frac{C^k x^{k-1} e^{-Cx}}{\Gamma(k)}
$$

—the Exponential and Gamma Distributions—we take $\rho_F(F/g) = \exp[-F/g]$.

The physically reasonable solutions are, again respectively,

$$g = -\frac{F}{2\,W\left(m, \frac{1}{2\sqrt{\frac{1}{F^2(R)}}}\right)}$$

$$L = \frac{1}{g}\exp[-C R_0/g]\,(C R_0 + g)$$

$$g = -\frac{F}{2\,W\left(m, \frac{1}{2\sqrt{\frac{C}{F^2 k}}}\right)}$$

$$L = \frac{\Gamma\left(k+1, \frac{C R_0}{g}\right)}{\Gamma(k)\,k} \tag{2.14}$$

where, again, $W(m, x)$ is the Lambert W-function, here taken of order $m = 0$.

Defining $F(\beta)$ as Eq. (2.10), and taking $C = R_0 = 1$, gives the cognition rate versus arousal relations of Fig. 2.3a and b. In Fig. 2.3b we allow the order of the Gamma Distribution to vary as an index of the 'depth' of cognitive layering.

A surprisingly stable form of the Yerkes-Dodson pattern persists, again bracketed by 'hallucination' and 'panic' modes, but without imaginary-valued components for the Exponential Distribution. The Gamma Distribution example, having a 'depth-of-decision' index k, displays sometimes nonzero imaginary valued components—the blue/lower level stratum.

For extreme value 'fat-tailed' distributions, on the other hand, cognition rate is indeterminate.

Much, then, depends on the probability model acting across the siloed enterprise limit. If individual components can go haywire, producing 'extreme events' with significant probabilities—perhaps akin to autoimmune disease within a higher organism—overall systemic cognition collapses.

A striking perceptual failure is, of course, the conclusion that some given distribution is 'well behaved' based on a limited set of observations. After all, crosstalk within an institution or other cognitive system provides regulatory feedback to limit or prevent down-the-rabbit hole excursions of important subcomponents. Failure of regulatory crosstalk, in this model, is like lighting a fuse to a bomb.

2.5 Fog and Friction

Direct

The development leading to Eq. (2.5) can, more or less easily, be extended in a standard manner to the imposition of stochastic burden on the systems characterized by Eqs. (2.7) and (2.14). Then the relation $dZ/dt \sim f(Z)$ becomes the stochastic differential equation (Cyganowski et al. 2002)

$$dZ_t = f(Z_t)dt + \sigma h(Z_t)dB_t \tag{2.15}$$

where dB_t represents Brownian white noise and $h(Z)$ will be set to Z as 'ordinary' volatility. Then, using the Ito Chain Rule (again, Cyganowski et al. 2002), we can calculate the nonequilibrium steady state conditions $\langle dL_t \rangle = 0$, taking $f(Z) = \beta - \alpha Z$ set at the peak of Figs. 2.2a and 2.3, i.e., $\alpha = 1,\ \beta = 3,\ C_1 = -1,\ C_2 = 3,\ R_0 = 1$. Again, for the multi-stream system, $C = 1$.

In Fig. 2.4a we calculate $Z(\sigma^2)$ and in Fig. 2.4b $Z(\sigma^2, \langle R \rangle)$ for the two (markedly) different expressions of the principal branches for $\langle dL_t \rangle = 0$ that follow from the formulas for cognition rate L in Eqs. (2.7) and (2.14).

The principal mode for the single-stream exponential distribution model displays bifurcation and a degradation of $Z(\sigma^2)$ until a critical value for $\sigma^2 \approx 3$. By contrast, the siloed multi-stream system, while similarly degraded by rising 'noise', shows complicated and counterintuitive patterns depending on $\langle R \rangle$.

Again, Caveat Emptor.

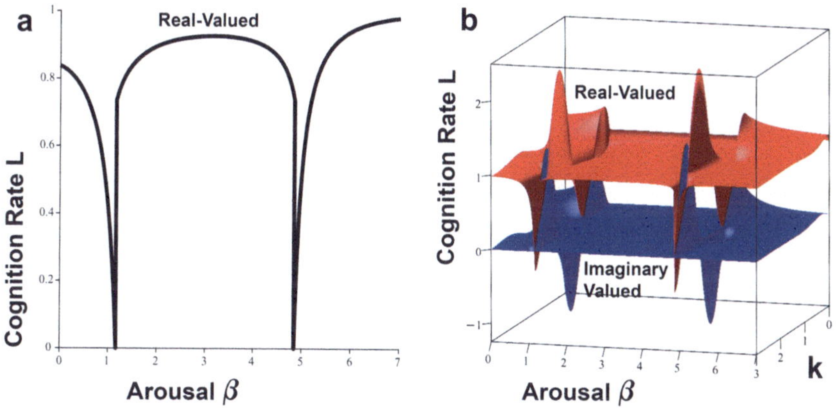

Fig. 2.3 Analog to Fig. 2.2 for solvable multi-stream models. **a** Based on the Exponential and Boltzmann Distributions. **b** Based the Gamma and Boltzmann Distributions. A form of the Yerkes-Dodson pattern persists, again bracketed by 'hallucination' and 'panic', but without imaginary-valued components in the Exponential example. The Gamma Distribution model interleaves real-valued (red) and imaginary-valued (blue) components across hallucination and panic modes, depending on the 'depth-of-decision' index k. For extreme value 'fat-tailed' distributions without first-moments, cognition rate is indeterminate

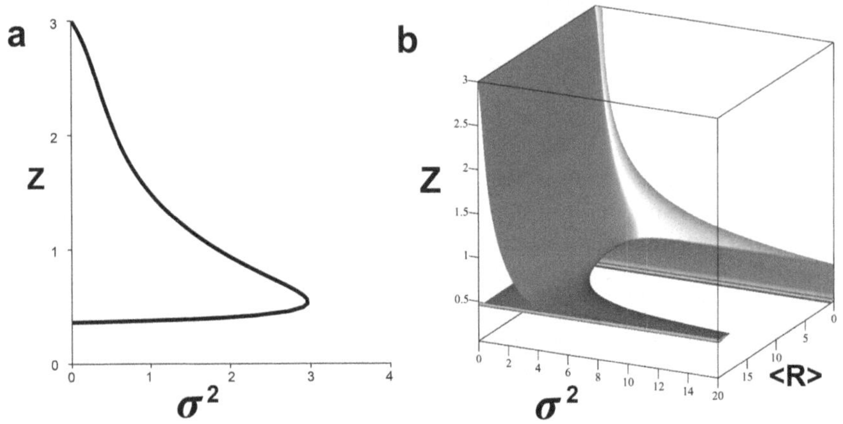

Fig. 2.4 **a** The principal mode for the single-stream exponential distribution model displays bifurcation and a degradation of $Z(\sigma^2)$ until a critical value for $\sigma^2 \approx 3$. **b** The siloed multi-stream system, while similarly degraded by rising 'noise', displays a complicated and counterintuitive dependence on $\langle R \rangle$

Delayed

Previously, α in the expression $dZ/dt = \beta - \alpha Z(t)$ simply characterized a kind of 'friction'. This will likely be compounded with delay, here assumed to be a fixed quantity τ, so that $dZ/dt = \beta - \alpha Z(t - \tau)$. Imposing

$$Z(t) \equiv \frac{\beta}{\alpha}(1 - \exp[\mathcal{S} t]) \tag{2.16}$$

gives

$$\begin{aligned} \mathcal{S} &= \frac{W(n, -\alpha\tau)}{\tau} \\ dZ/dt &= \mathcal{S}(Z - \beta/\alpha) \\ dZ_t &= \mathcal{S}(Z_t - \beta/\alpha)\,dt + \sigma Z_t dB_t \end{aligned} \tag{2.17}$$

$W(n, x)$ is again the n-th order Lambert W-function, assuming here $n = 0$, so that the synergism of 'friction' α with delay τ imposes a sharp phase transition when $\alpha\tau > \exp[-1]$.

We recalculate the nonequilibrium steady state conditions for cognition rates, $\langle dL_t \rangle = 0$, as in Fig. 2.4, fixing $\langle R \rangle = 1$, to give Fig. 2.5a and b. Note the phase transition to collapse for both at critical delay, in addition to that imposed by rising 'noise' σ^2.

Wallace (2022) gives further details. Extension to distributed delay requires other methods.

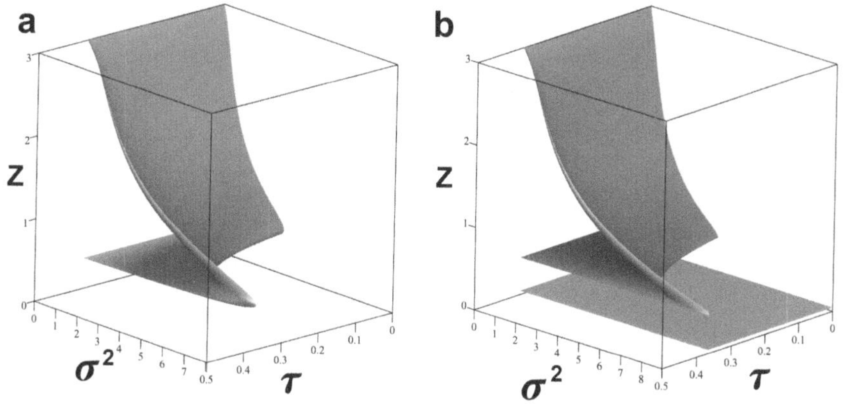

Fig. 2.5 We recalculate the nonequilibrium steady state conditions for the cognition rates $\langle dL_t \rangle = 0$ as in Fig. 2.4, setting $\langle R \rangle = 1$. **a** Context-driven system. **b** Salient object system. Since $\alpha = 1$ in both cases, the system fails as a phase transition at $\tau = \exp[-1] \approx 0.37$

Constrained

This argument can be further extended by recognizing that a 'system' is usually built from many interacting subcomponents that must operate under limited resources and time-to-act.

A relatively simple formal address can be made via 'ordinary' Lagrangian optimization.

We optimize cognition rates L_i of subcomponents under constraints of resource rates and time $Z = \sum_i Z_i$, $T = \sum_i T_i$ under the Lagrangian expressions

$$
\begin{aligned}
\mathscr{L} = \sum_i L_i + \lambda \left(Z - \sum_i Z_i \right) + \mu \left(T - \sum_i T_i \right) \\
\partial \mathscr{L} / \partial Z = -\lambda, \ \partial \mathscr{L} / \partial T = -\mu \\
\partial \mathscr{L} / \partial Z_i = \lambda, \ \partial \mathscr{L} / \partial T_i = \mu
\end{aligned}
\tag{2.18}
$$

λ and μ are *economic shadow prices* (Jin et al. 2008; Robinson 1993) representing the burden of additional environmental constraints.

Setting $dZ/dT = f(Z)$,

$$
\begin{aligned}
dZ/dT = \omega \equiv \frac{\mu}{\lambda} = f(Z) \\
Z(T) = f^{-1}(\omega)
\end{aligned}
\tag{2.19}
$$

Taking $f(Z) = \beta - \alpha Z$ or $\mathscr{S}(Z - \beta/\alpha)$, respectively,

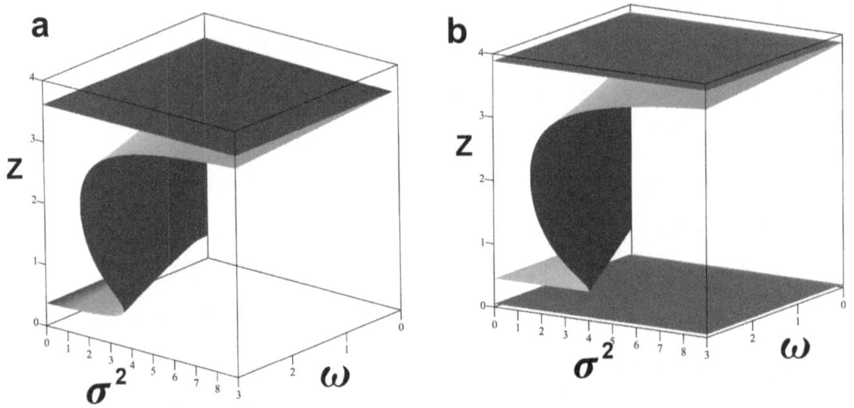

Fig. 2.6 Analog to Fig. 2.4, taking $\langle R \rangle = 1$, given resource and time constraints under shadow price burdens. **a** Context-Driven system. **b** Salient Objects system. Both principal branches show punctuated transitions under sufficient burden of 'fog' σ. The upper sections of the curved surfaces will not be reached under resource limitations

$$Z = \frac{\beta - \omega}{\alpha}$$
$$Z = \frac{\beta}{\alpha} + \frac{\omega}{\mathcal{S}} \tag{2.20}$$

remembering that $\mathcal{S}$ from the delay calculation will be negative if real-valued and that $Z_i \leq Z$.

As ω rises, resources collapse.

Because $dZ/dT = \omega$, it is possible to reconsider the stochastic models above, condensing Eq. (2.15) as

$$dZ_t = \omega dt + \sigma Z_t dB_t \tag{2.21}$$

Taking the same conditions leading to Fig. 2.4 now leads to Fig. 2.6, where $\langle R \rangle = 1$.

The primary branches—the curved sections—display punctuated transitions at sufficient levels of 'fog' σ. The upper limits of the bifurcated surfaces cannot be reached under constraints on total Z.

The Appendix reconsiders constraint on resources delivered to a multicomponent system from the perspective of a Shannon-Hartley reconfiguration of the model of Fig. 2.1. Dynamics are found to be driven by a synergism between shadow price ratio, the time constant of embedding environmental events, and the noise inherent to the overall control loop.

2.6 Intermediate Models

The fundamental difference for the two limits studied here—unsiloed versus siloed—is the contrast between Eqs. (2.2) and (2.12). A more general model can be constructed assuming

$$dP \sim \frac{Q(R)\rho_R(C,\ R/g)dR}{\int_0^\infty Q(R)\rho_R(C,\ R/g)dR} \tag{2.22}$$

where the function $Q(R)$ determines and drives the overall dynamics, replacing Eqs. (2.3), (2.4) and (2.13) with the obvious generalizations.

Taking $\rho_F = \exp[-F/g]$, $\rho_R = C\exp[-CR/g]$, with $Q(R) = R^n$,

$$\begin{aligned} g &= -\frac{F}{(n+1)\,W\left(m,\ -\frac{FC^{-\frac{n}{n+1}}\,\Gamma(n)^{\frac{1}{n+1}}\,n^{\frac{1}{n+1}}}{n+1}\right)} \\ L &= 1 - \frac{g^{-\frac{n}{2}}\,C^{\frac{n}{2}}\,R_0^{\frac{n}{2}}\,\mathrm{e}^{-\frac{CR_0}{2g}}\,M_{\frac{n}{2},\frac{n}{2}+\frac{1}{2}}\left(\frac{CR_0}{g}\right)}{\Gamma(2+n)} \end{aligned} \tag{2.23}$$

$W(m,\ x)$ is again the Lambert W-function of order m, and M is the Whittaker M-function of proper order.

Taking F as in Eq. (2.10), setting $\alpha = 1,\ C_1 = -1,\ C_2 = 3$, Fig. 2.7 displays the cognition rate L as a function of arousal β and n for $Q(R) = R^n$. There are two modes, representing the $m = 0,\ -1$ orders of the Lambert W-function. The results are familiar. For the general case, however, much will depend on the character of the distribution ρ_R.

A model based on the second-order Erlang Distribution, representing a 'two-stage exponential decision process', gives something similar to Fig. 2.7.

For the Cauchy distribution

$$\rho_R(C,\ R) = \frac{2}{\pi C\left(1 + (R/C)^2\right)}$$

that is without finite first moment, again taking $\rho_F = \exp[-F/g]$ and $Q(R) = R^n$,

$$\begin{aligned} \mathrm{e}^{-\frac{F}{g}} &= \int_0^\infty R^n\rho_R(C,\ R/g)dR = g^{n+1}C^n\sec\left(\frac{\pi n}{2}\right) \\ g &= -\frac{F}{(n+1)\,W\left(m,\ -\frac{FC^{\frac{n}{n+1}}\cos\left(\frac{\pi n}{2}\right)^{-\frac{1}{n+1}}}{n+1}\right)} \end{aligned}$$

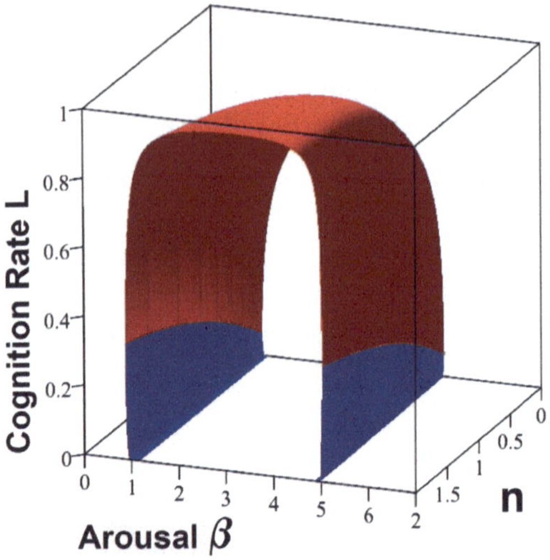

Fig. 2.7 Cognition rate $L(\beta, n)$ for the model of Eq. (2.23), setting $C = R_0 = 1$, $m = 0$. For the W-function, $m = 0$ is the upper sector, and $m = -1$ the lower. The second-order Erlang Distribution model gives something similar

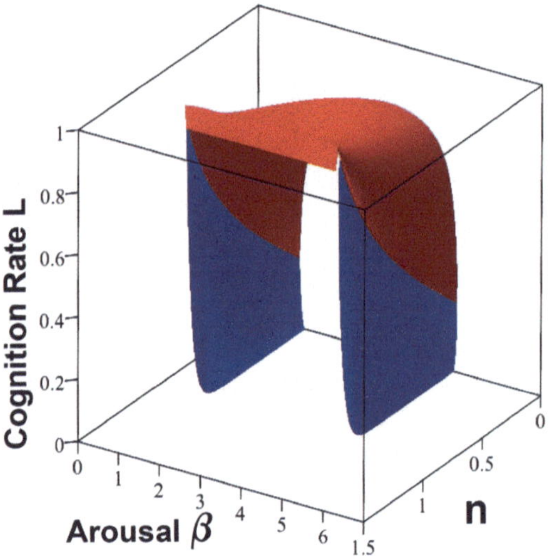

Fig. 2.8 Analog to Fig. 2.7 for the fat-tailed Cauchy distribution. Cognition collapses precipitously at $n = 1$ in the transition from unsiloed to siloed modes. Again, the top and bottom modes represent $m = 0, -1$ in the W-function expression

$$L = \frac{C^{1-n} g^{1-n} R_0^{n-1} \cos\left(\frac{\pi n}{2}\right) \Phi\left(-\frac{g^2 C^2}{R_0^2}, 1, \frac{1}{2} - \frac{n}{2}\right)}{\pi} \tag{2.24}$$

where Φ the Lerch Phi-function. Note that $\sec(\pi n/2)$ in the first expression is singular at $n = 1$.

Setting $C = R_0 = 1$ and assuming F as in Eq. (2.10) with $\alpha = 1$, $C_1 = -1$, $C_2 = 3$, gives Fig. 2.8. Here, the cognition rate collapses at $n = 1$ in the transition from unsiloed to siloed pattern. Again, there are two modes corresponding to $m = 0, -1$ in the Lambert W-function.

Calculations for the Levy Distribution—also without finite first moment—are left as an exercise. Note that the equivalent to the first expression in Eq. (2.24) then includes a multiplicative term in $\Gamma(1/2 - n)$, singular at $n = 1/2$ instead of at $n = 1$; matters can become even more intractable.

2.7 Discussion

We have examined in particular detail two simplified limiting cases of cognitive enterprise; a single-stream unsiloed system in which individual subcomponents are well-integrated into a definable whole, and a multi-stream fully-siloed case characterized by a probability distribution across the disjunct silos. While these examples provide contrasting insights in a first approximation, many real-world systems will likely be intermediate in structure and dynamics and may require more general methods, similar to those introduced above.

For many basic, underlying probability distribution examples—thin or fat-tailed—the well-integrated, single-stream system can generate a classic Yerkes-Dodson inverse-U pattern of cognition rate versus arousal, in the sense studied by Wallace and Fricchione (2024).

Well-behaved basic underlying probability distributions—having finite first moments—according to this analysis, can also provide a relatively stable foundation for the cognitive dynamics of independent, massively parallel, highly-siloed, multi-stream systems, even producing an inverse-U signal transduction with 'arousal', although still bracketed by some forms of 'hallucination' and 'panic'.

Stability in siloed systems, however, requires an absence of the 'rare events' that make up the fat-tails of distributions without well-defined first moments. Massively parallel siloed systems burdened with fat-tail distributions for internal process suffer complete cognitive collapse.

A recurring conundrum surrounds the identification of fat-tailed distributions from available data: One may need a very long-term data set, and in the absence of such, may misidentify an inherently fat-tailed system as otherwise.

For massively-parallel siloed enterprise—or even for the 'intermediate' case models of the previous section—erosion of regulatory systems that suppress—or at least mitigate—internal 'extreme events' would seem to characterize the synergistic burdens of stress and aging. Application to cognitive phenomena across different modes and various scales and levels of organization seems direct, including reconsideration of today's badly-overhyped AI entities that have little or no real-time 'embodied' interaction with their embedding environments.

Integrated systems of any nature, in which crosstalk between subelements is a matter of routine 'doctrine' so that there is an effective single large-scale cognitive stream, can remain relatively stable even in the presence of fat-tail excursions afflicting individual components. This being said, crosstalk between subentities of a larger cognitive structure will always be somewhat limited, in the further context of the inherent instability of some (most?) cognitive subdynamics, suggesting that the close pairing of cognition with internal regulation must always be a matter of degree. Fully-siloed systems, we would argue, remain in need of literally impossible levels of suppression of extreme events, of either endogenous or exogenous origin, remembering that relatively short-term observation often masks critical long-term dynamics.

Adaptation of these methods to intermediate structures has been broadly outlined in the previous section, and further development in this direction might lead toward statistical tools useful in the study and remediation of the failure dynamics studied here at and across a broad spectrum of modes, scales, and levels of organization.

2.8 Appendix

We reconsider the 'constrained' multicomponent system of Eq. (2.18) from the perspective a Shannon-Hartley model applied to Rate Distortion Control Theory model of Fig. 2.1. That is, Fig. 2.1 is now envisioned as transmission of a 'message' between a controlling agency and the system being controlled, in the presence of distorting 'topological information' and 'noise'.

The corresponding Shannon-Hartley channel capacity C is then

$$C = B \ln\left(1 + \frac{ZT/\tau}{S^2}\right) \tag{2.25}$$

where B is the channel 'bandwidth', Z and T are the totals as in Eq. (2.18), τ is a characteristic time constant of environmental effects, and S^2 is the in-channel noise. The expression T/τ ensures dimensional consistency, since Z is the rate variate necessary for the S-H calculation.

By Eq. (2.18), under shadow price constraints, $dZ/dT = \omega$, $T = Z/\omega$, so that Eq. (2.29) becomes

$$C = B \ln\left(1 + \frac{Z^2}{\omega\tau S^2}\right) \tag{2.26}$$

again in the context of the base SDE of Eq. (2.21).

Using the Ito Chain Rule to calculate the nonequilibrium steady state condition $\langle dC_t \rangle = 0$ from Eq. (2.30), some algebra gives

$$\sigma^2 = \frac{2\omega}{Z} \frac{\left(\omega\tau S^2 + Z^2\right)}{\left(Z^2 - \omega\tau S^2\right)} \tag{2.27}$$

Since $\sigma^2 \geq 0$, provided $\omega > 0$, a necessary condition for system stability is

$$Z > \sqrt{\omega\tau} S \tag{2.28}$$

driven by the synergism of ‘control noise’ S with the environmental characteristics ω and τ. Sufficient ‘environmental burden’ interacting with inherent control channel noise will obviate any possible value of resource rate Z. This outcome can, perhaps, be viewed as yet another form of the Data Rate Theorem.

Note that, formally, if the shadow price ratio $\omega < 0$, the necessary condition becomes

$$Z < \sqrt{|\omega|\tau S} \tag{2.29}$$

strongly capping the stable performance range by structurally limiting the resource rate Z ‘manageable’ under load. Such scenarios may, in fact, be possible, i.e., ‘fuse-blowing’, particularly in the case of a ‘rapid’ environment having small characteristic time constant τ, perhaps consonant with John Boyd’s famous ‘OODA Loop’ perspective.

References

Atlan, H., Cohen, I.: Immune information, self-organization, and meaning. Int. Immunol. **10**, 711–717 (1998)

Barabasi, A., Albert, R.: Emergence of scaling in random networks. Science **286**, 509–512 (1999)

Bennett, C.H.: The thermodynamics of computation. Int. J. Theor. Phys. **21**, 905–940 (1982)

Brown, R.: From groups to groupoids: a brief survey. Bull. Lond. Math. Soc. **19**, 113–134 (1987)

Champagnat, N., Ferriere, R., Meleard, S.: Unifying evolutionary dynamics: from individual stochastic process to macroscopic models. Theor. Popul. Biol. **69**, 297–321 (2006)

Cover, T., Thomas, J.: Elements of Information Theory, 2nd edn. Wiley, New York (2006)

Cyganowski, S., Kloeden, P., Ombach, J.: From Elementary Probability to Stochastic Differential Equations with MAPLE. Springer, New York (2002)

de Groot, S., Mazur, P.: Non-Equilibrium Thermodynamics. Dover, New York (1984)

Dembo, A., Zeitouni, O.: Large Deviations and Applications, 2nd edn. Springer, New York (1998)

Derman, E., Miller, M., Park, D.: The Volatility Smile. Wiley, New York (2016)

Dolan, B., Janke, W., Johnston, D., Stathakopoulos, M.: Thin fisher zeros. J. Phys. A **34**, 6211–6223 (2001)

Dretske, F.: The explanatory role of information. Phil. Trans. R. Soc. A **349**, 59–70 (1994)

Feynman, R.: Lectures on Computation. Westview Press, New York (2000)

Fisher, M.: Lectures in Theoretical Physics, vol. 7. University of Colorado Press, Boulder (1965)

Genewein, T., Leibfried, F., Grau-Moya, J., Braun, D.: Bounded rationality, abstraction, and hierarchical decision-making: an information-theoretic optimality principle, frontiers in Robotics and AI (2015). https://doi.org/10.3389/frobt.2015.00027

Harush, U., Barzel, B.: Dynamic patterns of information flow in complex networks. Nat. Commun. **8**, 2181 (2017)

Jackson, D., Kempf, A., Morales, A.: A robust generalization of the Legendre transform for QFT. J. Phys. A **50**, 225201 (2017)

Khinchin, A.: The Mathematical Foundations of Information Theory. Dover, New York (1957)

Laidler, K.: Chemical Kinetics, 3rd edn. Harper and Row, New York (1987)

Landau, L., Lifshitz, E.: Statistical Physics, 3rd edn, Part 1. Elsevier, New York (2007)

Maturana, H., Varela, F.: Autopoiesis and Cognition: The Realization of the Living. Reidel, Boston (1980)

Mezo I., Keady, G.: Some physical applications of generalized Lambert functions, arXiv:1505.01555v2 [math.CA] 22 Jun 2015 (2015)
Nair, G., Fagnani, F., Zampieri, S., Evans, R.: Feedback control under data rate constraints: an overview. Proc. IEEE **95**, 108138 (2007)
Pettini, M.: Geometry and Topology in Hamiltonian Dynamics and Statistical Mechanics. Springer, New York (2007)
Ruelle, D.: Cluster property of the correlation functions of classical gases. Rev. Modern Phys. 580–584 (1964)
Taleb, N.N. (ed.).: Statistical Consequences of Fat Tails: Real World Preasymptotics, Epistemology, and Applications. STEM Academic Press (2020)
Wallace, R., Fricchione, G.: Stress-induced failure of embodied cognition: a general model. BioSystems **239**, 105193 (2024). https://www.sciencedirect.com/science/article/abs/pii/S0303264724000789
Wallace, R.: Consciousness: A Mathematical Treatment of the Global Neuronal Workspace Model. Springer, New York (2005)
Wallace, R.: Consciousness, crosstalk, and the mereological fallacy: an evolutionary perspective. Phys. Life Rev. **9**, 426–453 (2012)
Wallace, R.: Canonical Instabilities of Autonomous Vehicle Systems: The Unsettling Reality Behind the Dreams of Greed. Springer, New York (2018)
Wallace, R.: Cognitive Dynamics on Clausewitz Landscapes: The Control and Directed Evolution of Organized Conflict. Springer, New York (2020)
Wallace, R.: How AI founders on adversarial landscapes of fog and friction. J. Defense Model. Simul. (2021). https://doi.org/10.1177/1548512920962227
Wallace, R.: Consciousness, Cognition, and Crosstalk: The Evolutionary Exaptation of Nonergodic Groupoid Symmetry-Breaking. Springer, New York (2022)
Wallace, R.: On 'machine consciousness'. J. Artif. Intell. Consiousness **10**, 125–148 (2023a). https://doi.org/10.1142/S2705078522500187
Wallace, R.: Essays on the Extended Evolutionary Synthesis: Formalizations and Expansions. Sprnger, New York (2023b)
Wallace, R.: Mathematical Essays on Embodied Cognition: Insights from Information and Control Theories. Springer, New York (2025)
Watts, D., Strogatz, S.: Collective dynamics of small world networks. Nature **393**, 440–442 (1998)
Weinstein, A.: Groupoids: unifying internal and external symmetry. Notices Am. Mathe. Assoc. **43**, 744–752 (1996)
Yi, S., Nelson, P.W., Ulsoy, A.G.: Time-Delay Systems: Analysis and Control Using the Lambert W Function. World Scientific, New Jersey (2010)
Zilberstein, S.: Using anytime algorithms in intelligent systems. AI Mag. **17**, 73–83 (1996)

Chapter 3
In Praise of Mission Command and Workplace Autonomy

In general, complex models with many free parameters are likely to succeed in stable, stationary worlds, while simple heuristics, like human intelligence, evolved for dealing with an uncertain, social world...

—Gigerenzer (2021)

3.1 Introduction

'Western' studies of institutional form and function often focus on salient events involving individual 'managers', 'management strategy', and associated dynamics (e.g., Mintzberg 1973, 1989; Porter 1980, 2008). Other perspectives are possible (e.g., Gerstner 2003), viewing such salient events as entrained—indeed, entrapped—within larger-scale processes of institutional cognition that are themselves cultural artifacts embedded in path-dependent historical trajectories and their encompassing sociocultural streams. See (Wallace 2025) for worked-out examples that explore burdens imposed by multi-stage decision sequences on cognitive dynamics, given 'noise', time or resource constraints, and delay.

Here, we reconsider institutional cognitive dynamics under conditions of 'siloing,' in the context of underlying 'fat-tailed' probability distributions that greatly amplify the likelihood of extreme events.

The methods necessary for this analysis, while roughly analogous to the more quantitative 'bounded rationality' studies (e.g., Genewein et al. 2015), are firmly based on the asymptotic limit theorems of information and control theories, and this requires some detailed explication before proceeding to the central ideas.

© The Author(s), under exclusive license to Springer Nature Switzerland AG 2026

R. Wallace, *Bounded Rationality and Its Discontents*, Studies in Systems, Decision and Control 664, https://doi.org/10.1007/978-3-032-18423-8_3

3.2 A Review of Methodology

The Asymptotic Limit Theorems

A small set of asymptotic limit theorems from information and control theories constrains all possible cognitive phenomena, 'semantic' or otherwise (Cover and Thomas 2006; Dretske 1994; Khinchin 1957; Nair et al. 2007). Building new, robust, statistical tools for the analysis of institutional—and other—cognitive dynamics and dysfunctions depends on these theorems, just as regression models, t-tests, and so on, are based on the Central Limit Theorem, the Renewal Theorem, the Martingale Theorem, and friends.

The limit theorems of interest are:

- The Shannon Coding Theorem that prescribes how to code a message in 'typical' sequences for maximum error-free transmission through a particular channel at a maximum 'channel capacity' (and the 'tuning theorem' variant in which the message formally 'transmits' the channel).
- The Source Coding Theorem that splits a set of sequences emitted by an information source into two subsets, a relatively small one that is 'meaningful'—the sequences are consonant with an underlying 'grammar' and 'syntax'—and a much larger 'meaningless' one 'of measure zero'. Only for stationary, ergodic systems is it possible to characterize 'information' and 'information source uncertainty' in terms of a Shannon entropy-analog. For nonergodic systems, each (infinite) signal sequence has it's own scalar source uncertainty that cannot be expressed in 'Shannon entropy' form (Khinchin 1957). It is here that methods must be abducted from statistical physics, introducing as well equivalence-classes of sequences forming a groupoid, leading to groupoid symmetry-breaking phase transitions characterizing 'cognitive' phase transitions, quite unlike those of 'conventional' physical systems.
- The Rate Distortion Theorem (RDT) and an associated Rate Distortion Control Theory that, in a sense, 'tunes' the channel capacity around a maximum-permissible scalar measure of average distortion (which finesses the Friston 'free energy minimization' approach).
- The Data Rate Theorem (DRT) stating that, for stability of an inherently unstable control system, 'control information' must be provided at a rate greater than the system produces it's own 'topological information'. A version of this is, somewhat surprisingly, 'easily' derived from a time and 'free energy rate' optimization based on the Hartley-Shannon Theorem.

Further work shows:

- Cognition implies choice—of a small set of actions from a larger embedding set of what is possible to the system of interest. Such choice reduces uncertainty in a formal manner, implying the existence of an information source 'dual' to the cognitive process studied, at and across the varied scales and levels of organization of the living state (Atlan and Cohen 1998; Dretske 1994; Maturana and Varela 1980).

- Real-world, real-time cognition is always inherently unstable under the impact of the 'topological information' imposed by embedding environmental 'roadways'. The DRT implies that cognition must always be paired with regulation for system stability and long-term function under selection pressures. Indeed,, it can perhaps be argued that the degree of pairing between cognition and regulation measures a system's 'self awareness', a central feature of many consciousness models.
- We impose an adiabatically, piecewise, stationary (APS) 'Born Oppenheimer' model in which the system of interest remains 'close enough' to stationary during the period(s) of interest so that the asymptotic limit theorems are effectively 'anytime algorithms' (e.g., Zilberstein 1996). Between 'pieces' for which this assumption is sufficiently true, we invoke (groupoid) symmetry-breaking and/or Fisher Zero (Dolan et al. 2001) phase transitions.
- The ergodic restriction can be lifted, at the cost of imposing methods abducted from statistical physics, with dynamics approximated via an analog to the first (and higher) order Onsager treatment of nonequilibrium thermodynamics, recognizing, however, that, due to the failure of information source microreversibility—e.g., 'eht' does not have the same probability as 'the' in English—there are no 'Onsager reciprocal relations' in high order approximations.
- For purely physical phenomena, free energy is often defined in terms of Boltzmann distribution-derived partition functions. In contrast, free energy analogs associated with the dynamics of cognitive phenomena can, and may well, be characterized by 'fat-tailed' distributions that do not have well-defined first or second moments (e.g., Derman et al. 2016; Wallace and Fricchione 2024).

Resource Rates

Enterprises are embedded in an environment that includes themselves. There are at least three resource streams necessary for their proper function:

- The rate at which elements of the system can communicate with each other, instantiated by some channel capacity $\mathscr{C}$.
- The rate at which 'sensory information' is available from the embedding circumstance, say according to a channel capacity $\mathscr{H}$.
- The rate at which 'material/materiel' resources can be provided, a rate $\mathscr{M}$. For an organism, this might be measured by the rate at which metabolic free energy is provided. Machine/institutional systems will have other measures.

These rates will cross-correlate: one confronts a 3×3 (or larger) matrix $\mathbf{Z}$. An $n \times n$ matrix will have will have n scalar invariants r_i determined by the standard expression

$$\begin{aligned} p(\gamma) &= \det[\mathbf{Z} - \gamma \mathbf{I}] \\ &= (-1)^n \gamma^n + (-1)^{n-1} r_1 \gamma^{n-1} + \ldots - r_{n-1}\gamma + r_n \end{aligned} \tag{3.1}$$

I is the appropriate identity matrix, 'det' is the determinant, and γ is a real-valued parameter. The first invariant is the matrix trace, and the last the determinant. We seek to construct a scalar resource rate measure from these invariants.

The most direct such index might be $Z = \mathscr{C} \times \mathscr{H} \times \mathscr{M}$, but our worlds are not likely so simple.

We postulate—in first-order—that a single scalar index $Z(r_1, ..., r_n)$ can be constructed that sufficiently represents information and material resource rates. This approach is analogous to, but different from, the more usual principal component analysis based on a correlation matrix. Wallace (2021) outlines the complications associated with generalization to more than a single such index.

Rate Distortion Control Theory

A directly 'embodied' formalism emerges via a 'Rate Distortion Control Theory' based on the Rate Distortion Theorem. As described, the RDT addresses transmission under conditions of noise over an information channel. There will be, for that channel and a particular scalar measure of average distortion D between what is sent and what is received, a minimum necessary channel capacity $R(D)$.

The theorem asks, in effect, what is the 'best' channel for transmission of a message with the least possible average distortion. $R(D)$ can be defined for nonergodic information sources via a limit argument based on the ergodic decomposition of a nonergodic source into a 'sum' of ergodic sources. The RDT can be reconfigured in a control theory context, if we envision a system's topological information from the Data Rate Theorem as another form of noise, adding to the average distortion D. See Fig. 3.1.

The punctuation implied by the Data Rate Theorem (Nair et al. 2007) happens if there is a critical maximum average distortion beyond which control fails.

3.3 The First Model

We take a somewhat simplistic first-order approach, abducting standard formalism from statistical mechanics (Landau and Lifshitz 2007) that nonetheless requires a perhaps counterintuitive re-definition of 'temperature' for cognitive systems.

Consider the full ensemble of high probability developmental trajectories available to the system, written as Y_j, $j = 1, 2, ...$ Each trajectory is associated with a Rate Distortion Function-defined minimum necessary channel capacity R_j for a particular maximum average scalar distortion D_j. Then, assuming some basic underlying probability model having distribution $\rho_R(x)$, we can define a pseudoprobability for a meaningful 'message' Y_j sent into the system of Fig. 3.1 as

$$dP = \frac{\rho_R(R/g(Z))dR}{\int_0^\infty \rho_R(R/g(Z))dR} = \frac{\rho_R(R_j/g(Z))dR}{g(Z)} \tag{3.2}$$

since, by definition, $\rho_R(x)$ is a probability density function that integrates to one on the interval $x = 0 \rightarrow \infty$.

The 'temperature' $g(Z)$ in Eq. (3.2) deserves some comment. It is analogous to the inverse of the 'rationality parameter' of Bounded Rationality Theory (e.g., Genewein et al. 2015), but explicitly incorporates the resource rate index Z. Its functional form is not predetermined, but emerges from a first-principles calculation based on an iterated argument similar to Onsager's treatment of nonequilibrium thermodynamics. Recall that, according to Feynman (2000)—following Bennett (1982)—'information', and associated representations of channel capacity, are not 'entropies', but kinds of free energies from which an iterated actual entropy can then be defined. System dynamics arise from gradients of this 'real' entropy. Details will be given below for a first-order model. Higher order models can be imposed as empirically needed.

We only treat the continuous case. For discrete or finite-state systems, the integral is a sum, and the partition function denominator becomes some more general function $h(g(Z))$.

Again, Y_j is some particular trajectory, while the (generalized) integral is over all possible 'high probability' paths.

We—implicitly—impose the Shannon-McMillan Source Coding Theorem (Cover and Thomas 2006): the set of possible system trajectories can be divided in two. These are, first, a very large set of measure zero—vanishingly low probability—that is not consistent with the underlying grammar and syntax of some basic information source, and a much smaller consistent set (Khinchin 1957).

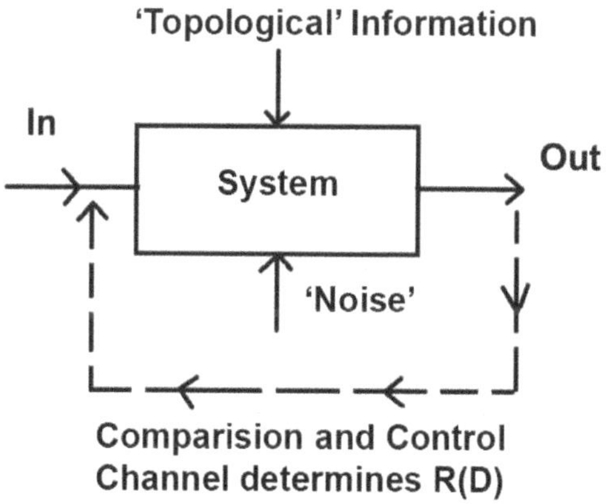

Fig. 3.1 The rate distortion control theory model. An inherently unstable system's 'topological information', according to the data rate theorem (Nair et al. 2007), is added to the effects of 'noise' to determine the minimum channel capacity $R(D)$ needed to assure transmission with average distortion D. The punctuation implied by the DRT emerges here if there is a critical maximum average distortion beyond which control fails

Again, R_j is the RDT channel capacity keeping average distortion less than a given limit D_j for message Y_j. $g(Z)$ is the temperature analog from above, depending on the scalar resource rate Z. $\rho(x)$, for physical systems, is usually taken as the Boltzmann distribution: $\rho(x) = \exp[-x]$. We suggest that, for cognitive phenomena—from the living and institutional to the machine and composite—it is necessary to move beyond analogs with physical systems. That is, it becomes necessary explore the influence of a variety of different probability distributions, including those with 'fat tails' (Derman et al. 2016; Taleb 2020), on the dynamics of cognition/regulation stability.

Again, standard methodology from statistical physics (Landau and Lifshitz 2007) identifies the denominator of Eq. (3.2) as a partition function, giving an iterated free energy-analog F as

$$
\begin{aligned}
\rho_F(F/g(Z)) &\equiv \int_0^\infty \rho_R(R/g(Z))dR = g(Z) \\
g(Z) &= \text{RootOf}\,(X - \rho_F(F/X))
\end{aligned}
\tag{3.3}
$$

where 'RootOf' means setting the expression to zero and solving for X and ρ_F is an appropriate 'free energy' distribution that must be determined case-by-case for cognitive, as opposed to simple physical, phenomena. For physical phenomena, ρ_F can often be taken as the Boltzmann distribution.

Using a standard argument from chemical physics (Laidler 1987), we define the cognition rate of the system as the reaction rate analog

$$
L \equiv \frac{\int_{R_0}^\infty \rho_R(R/g(Z))dR}{\int_0^\infty \rho_R(R/g(Z))dR} = \frac{1}{g(Z)} \int_{R_0}^\infty \rho_R(R/g(Z))dR \tag{3.4}
$$

R_0 is the minimum channel capacity needed to keep the average distortion below a critical value D_0 for the full system of Fig. 3.1.

The basic underlying probability model of Eq. (3.2)—via the partition function argument of Eq. (3.3)—determines system dynamics, but not system structure, and cannot be associated with a particular underlying network conformation, although these are related (Barabasi and Albert 1999). More specifically, a set of markedly different networks may all be mapped on to any single given dynamic behavior pattern and, indeed, vice-versa, The same static network may display a spectrum of behaviors (Harush and Barzel 2017). We focus on dynamics rather than network topology, a shift from much current work.

Feynman (2000)—following Bennett (1982)—identifies 'information' as a form of free energy, using Bennett's ideal machine that turns a message directly into work. We are concerned here with an iterated, rather than a direct, construction.

F, as a free energy, is subject to symmetry-breaking transitions as $g(Z)$ varies (Pettini 2007). These symmetry changes, however, are not as associated with physical phase transitions as represented by standard group algebras. Such symmetry changes represent transitions from playing one 'game' to playing another. For example, a cognitive system may engage in foraging behaviors that trigger a predatory attack by another system. Then the game changes from 'foraging' to 'escape'.

Thus 'cognitive phase changes' involve shifts between equivalence classes of high probability developmental/behavioral pathways that are represented as *groupoids*, a generalization of the group concept such that a product is not necessarily defined for every possible element pair, although multiple products with multiple identity elements are defined (Brown 1987; Weinstein 1996).

Dynamics emerge via a first-order Onsager approximation akin to that of nonequilibrium thermodynamics (de Groot and Mazur 1984) in the gradient of an entropy measure S constructed from the 'iterated free energy' F of Eq. (3.3) by means of a Legendre transform that should be familiar from the Black-Scholes approximation (Landau and Lifshitz 2007; Pettini 2007).

Recall from the (first-order) Onsager approximation of nonequilibrium thermodynamics that $\partial Z/\partial t \propto \partial S/\partial Z$. The full—again, standard, first-order—development is

$$
\begin{aligned}
&S(Z) \equiv -F(Z) + ZdF(Z)/dZ \\
&\partial Z/\partial t \approx dS/dZ = f(Z) \\
&f(Z) = Zd^2F/dZ^2 \\
&F(Z) = \int\int \frac{f(Z)}{Z} dZdZ + C_1Z + C_2 \\
&g(Z) = \text{RootOf}\,(X - \rho_F(C_F,\ F(Z)/X))
\end{aligned}
\tag{3.5}
$$

An expression for $g(Z)$ follows from the second expression of Eq. (3.3). Note that the 'diffusion coefficient' has been implicitly set equal to 1 in the second part of Eq. (3.5).

Again, higher order approximations are in the same realm as higher order regression models, modulo dimensional consistency. Full treatment may involve generalization of the Legendre transform via formal power series methods (Jackson et al. 2017).

Several points:

- Since 'RootOf' may have complex number solutions, the temperature analog $g(Z)$ enters the realm of the 'Fisher Zeros' characterizing phase transition in physical systems (e.g., Dolan et al. 2001; Fisher 1965; Ruelle 1964 Sec. 5). Such phase transitions lie at the base of the punctuated accession to consciousness in the Baars, and related, Global Workspace Models. Indeed, the 'RootOf' construction in Eq. (3.3) actually generalizes the Lambert W-function (e.g., Yi et al. 2010; Mezo and Keady 2015). It is the W-function and its generalizations that drive the inverse-U 'Yerkes-Dodson' stress responses.
- To reiterate, information sources are not microreversible, that is, palindromes are highly improbable, for example the sequence 'ti' has far lower probability than 'it'

in English. Thus, there are no 'Onsager Reciprocal Relations' in higher dimensional systems. The imposition of groupoid symmetries on cognitive phenomena appears to follow upon this directed homotopy.

Further, there will always be a delay in the rate of provision of Z, so that, in Eq. (3.5), for example, $f(Z) = \beta - \alpha Z(t)$—an exponential model having $Z(t) = (\beta/\alpha)(1 - \exp[-\alpha t])$—where $Z \to \beta/\alpha$ at a rate determined by α. Other dynamics are possible, such as the 'Arrhenius', $Z(t) = \beta \exp[-\alpha/t]$, with $f(Z) = (Z/\alpha)(\log(Z/\beta))^2$. However Z is constructed from the components $\mathscr{C}$, $\mathscr{H}$ and $\mathscr{M}$, so here, it is the scalar resource rate Z itself that concerns us.

Taking $f(Z) = \beta - \alpha Z$, the third expression in Eq. (3.5) leads to an explicit expression for F as

$$F = \beta(Z \ln(Z) - Z) - \frac{\alpha Z^2}{2} + C_1 Z + C_2 \tag{3.6}$$

3.4 The General Model

Systems based in 'Western' cultural ideology/biology will focus on collections of salient events almost always at the exclusion of important contextual drivers. Wallace (2020) follows others in exploring the consequences of this systematic omission on important collective enterprise.

For a context-free 'singular events' system, the most elementary approach is both absurdly simple and deeply subtle.

Imagine a very large parallel set of dual information sources having independent channel capacities $R_i(D_i)$ and define a probability distribution across the ensemble that replaces Eq. (3.2), along with a cognition rate that replaces Eq. (3.4), as

$$\begin{aligned} P_j &= \frac{R_j}{\sum_k R_k} = \frac{R_j}{R} \\ L &= \frac{1}{R} \sum_{R_j > R_0} R_j \end{aligned} \tag{3.7}$$

In the limit of many possible channels and 'intermediate' structures, this generalizes as

$$dP \sim \frac{R^n \rho_R(R/g) dR}{\int_0^\infty R^n \rho_R(R/g) dR} = \frac{R^n \rho_R(R/g) dR}{g^2 \langle R^n \rangle} \tag{3.8}$$

where $\rho_R(x)$ is the fundamental, underlying probability distribution across the system of concern. $\langle R^n \rangle$ is the mean value of R^n, if it indeed exists. For the fully-integrated model, $n = 0$. For the fully-siloed model, $n = 1$. In general, we take $0 \leq n \leq 1$.

The distribution ρ_R across the world of singular events must have a finite first moment for a fully-siloed system to function, a matter that has obvious implications for the cognition rate. In particular, $g^2 \langle R^n \rangle$ becomes the partition function in a further iteration, where the relations of interest are

$$
\begin{aligned}
\rho_R(F/g) &= g^2 \left\langle R^n \right\rangle \\
L &= \frac{\int_{R_0}^{\infty} R^n \rho_R(R/g)\, dR}{\int_0^{\infty} R^n \rho_R(R/g)\, dR} = \frac{\int_{R_0}^{\infty} R^n \rho_R(R/g)\, dR}{g^2 \langle R^n \rangle}
\end{aligned} \tag{3.9}
$$

$\rho_R(F/g)$ is taken as representing an empirically-appropriate free energy distribution that—here—we assume is the same as ρ_R, and R_0 is the function-threshold channel capacity.

As above, the first expression can be solved for the 'temperature' parameter g in in terms of the free energy analog F, and this inserted into the expression for cognition rate L to define a function $L(\beta)$, assuming Eq. (3.6).

3.5 'Simple' Multi-stage Systems

If one takes the basic, underlying probability model to be an Exponential Distribution, $\rho(x) = C \exp[-Cx]$, then the Erlang Distribution of order k represents the sum of k independent, identically distributed exponential random variables. We assume the exponential time dynamics in resource flow leading to Eq. (3.6), taking $\alpha = 1$, $C_1 = -1$, $C_2 = 3$, and, since $Z(t) \to \beta/\alpha$, define β as the index of 'arousal'.

Here, we take k, as n above, to be a real number $k \geq 1$. Then the Erlang Distribution of order k-positive integer becomes the Gamma Distribution of order $k \geq 1$ real and the expressions for 'temperature' g and the rate of cognition L become, after some tedious algebra,

$$
\begin{aligned}
\rho(C, k, x) &= \frac{C^k x^{k-1} e^{-Cx}}{\Gamma(k)} \\
g &= -\frac{FC}{(n+k)\, W\left(m, -\frac{F\left(F^{k-1}\right)^{-\frac{1}{n+k}} \Gamma(n+k)^{\frac{1}{n+k}}}{n+k}\right)} \\
L &= \frac{\Gamma\left(n+k, \frac{CR_0}{g}\right)}{\Gamma(n+k)}
\end{aligned} \tag{3.10}
$$

where $W(m, x)$ is the Lambert W-function that satisfies the relation

$$
W(m, x) \exp[W(m, x)] = x
$$

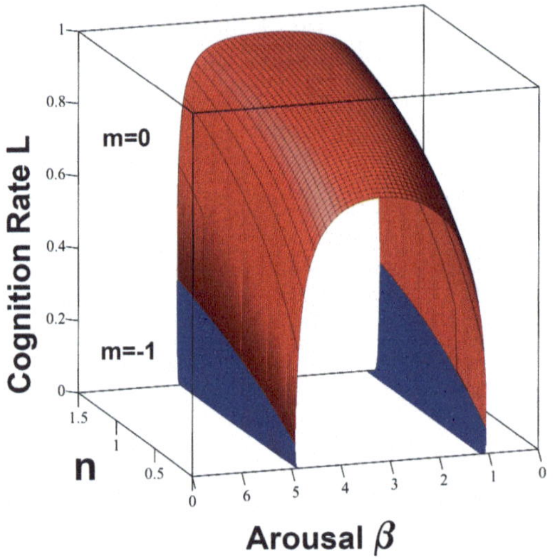

Fig. 3.2 Cognition rate versus 'arousal' for the $k = 1$ Gamma/exponential distribution model across variation in the silo-index n. For an integrated, one-step 'mission command' cognition process, even siloed systems can follow a classic Yerkes-Dodson inverse-U with arousal

It is real-valued only for $m = 0, -1$ and $x \geq -\exp[-1]$, introducing highly-punctuated phase transition dynamics.

We will take $C = R_0 = 1$. The mean value for the Gamma Distribution is simply k/C: It is not at all fat-tailed, but system cognitive dynamics are indeed dependent on the n-exponent as well as the order of the decision-process, k.

We first suppose that the basic underlying probability distribution is the Boltzmann, i.e., the Gamma Distribution with $C = k = 1$, and, in Fig. 3.2, allow the siloing processes for cognition rate to range across $0 \leq n \leq 1.5$. The two colors represent the two Lambert W-function solutions with $m = 0, -1$.

A classic Yerkes-Dodson inverse-U pattern with 'arousal' emerges (e.g., Wallace and Fricchione 2024).

Next, in Fig. 3.3, we fix the degree of siloing as $n = 0, 1$, and allow k, the index of decision-level depth micromanagement to vary. A different pattern emerges. Only for $k = 1$—equivalent to the military doctrine of 'mission command'—does this system follow the classic Yerkes-Dodson inverse-U (the very thin lines at $k = 1$). $k > 1$ micromanagement systems, in a sense, fall off a cliff regardless of siloing level.

The Appendix explores conformations and contortions of the system represented in Fig. 3.2 under stochastic burden, as taken at the peak of that figure, i.e., $\alpha = 1$, $\beta = 3$ for $dZ/dt = f(Z) = \beta - \alpha Z$.

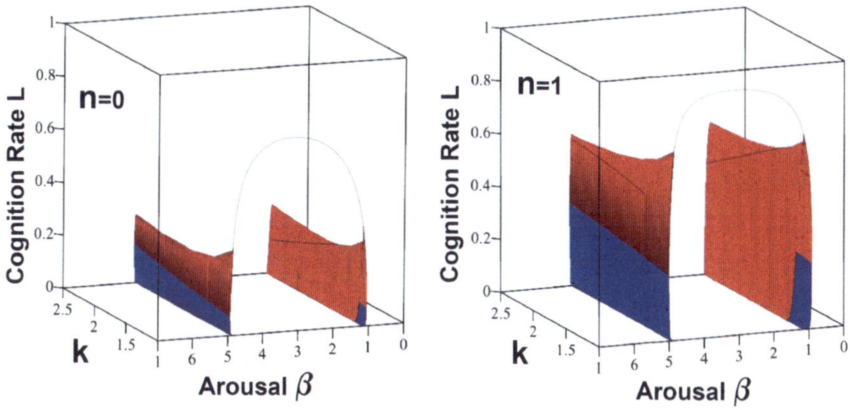

Fig. 3.3 Fixing siloing at $n = 0, 1$, k increases as an inverse degree of 'mission command'. Only for $k = 1$ (the thin lines), do these systems follow the classic inverse-U of cognition rate with arousal

3.6 Extreme Events

We next explore the impact of siloing on systems characterized by three 'fat-tailed' distributions, dominated by extreme events and without well-defined mean value first moments. In general, these systems do not have exact algebraic solutions for cognition rate and must be addressed numerically. Here, we will study patterns in the cognition temperatures rather than the cognition rates. The presence or absence of inverse-U cognition rates with arousal is closely mirrored by cognition temperature. That is, $L(Z)$ has the same topology as $g(Z)$.

We set $F(Z)$ as in Eq. (3.6), again taking $\alpha = 1$, $C_1 = -1$, $C_2 = 3$ and letting β vary as the index of arousal.

Recall that, for a distribution ρ_R, given F, g can be solved for from

$$\rho_R(F/g) = \int_0^\infty R^n \rho_R(R/g) dR \tag{3.11}$$

The Cauchy, $1/(1+x)^2$, and Levy Distributions, which are all without first moments, generate the relations

$$\frac{2}{\pi C\left(1+\frac{F^2}{g^2C^2}\right)} = g^{n+1}C^n \sec\left(\frac{\pi n}{2}\right)$$

$$\frac{1}{\left(\frac{F}{g}+1\right)^2} = g^{n+1}\pi n \csc(\pi n)$$

$$\frac{\sqrt{2}\sqrt{\frac{C}{\pi}}\,e^{-\frac{Cg}{2F}}}{2\left(\frac{F}{g}\right)^{3/2}} = \frac{C^n 2^{-n} g^{n+1}\Gamma\left(\frac{1}{2}-n\right)}{\sqrt{\pi}} \tag{3.12}$$

We take $C = 1$, and use the Implicitplot3d function of the computer algebra program MAPLE 2025 to calculate the various forms of $g(n, \beta)$ in Fig. 3.4.

The Cauchy (**a**) and Levy (**c**) distributions are pathological at all values of n. The nameless $1/(1+x)^2$ distribution, however, shows two classic inverse-U solutions that collapse at the $n = 1$ full-silo limit. That distribution has the simple monotonic-declining hazard rate $1/(1+x)$, while the others have quite involved hazard rates.

Notice that the Levy Distribution example, by virtue of the Gamma Function term, becomes singular at $n = 1/2$.

The Levy Distribution is a special case of the Inverse Gamma Distribution having the fundamental relation

$$\frac{g^{k+1}C^k\left(\frac{1}{F}\right)^k e^{-\frac{Cg}{F}}}{F\Gamma(k)} = \frac{C^n g^{1+n}\Gamma(-n+k)}{\Gamma(k)} \tag{3.13}$$

which is singular at $n = k$. For the Levy Distribution, $C = k = 1/2$.

3.7 Discussion

As expected, Eqs. (3.8) and (3.9) imply that fully-siloed enterprises—having sufficiently-large siloization index n—that confront environments dominated by sufficiently fat-tailed distributions of extreme events simply cannot function under such burden.

In some measure, however, the fat-tailed Nameless Distribution, $\rho_R(C, x) = C/(1+Cx)^2$, with its simple monotonic decreasing hazard rate, does indeed follow the Yerkes-Dodson effect up to a critical level of siloization. This suggests that an institution might, in theory, 'tune' it's internal structural and environmental response doctrines to, in some measure, compensate for the occurrence of extreme events, short of complete internal siloization.

In this context, however, the 'simple' $k = 1$ single-stage Exponential Distribution cognition model of Sect. 3.5 and Fig. 3.2 provides something of a surprise. A $k = 1$ Exponential Distribution model, in this formalism, represents a Mission Control/Workplace Empowerment institutional doctrine in which even the most isolated subcomponent is free to interpret higher-level orders according to the demands of its

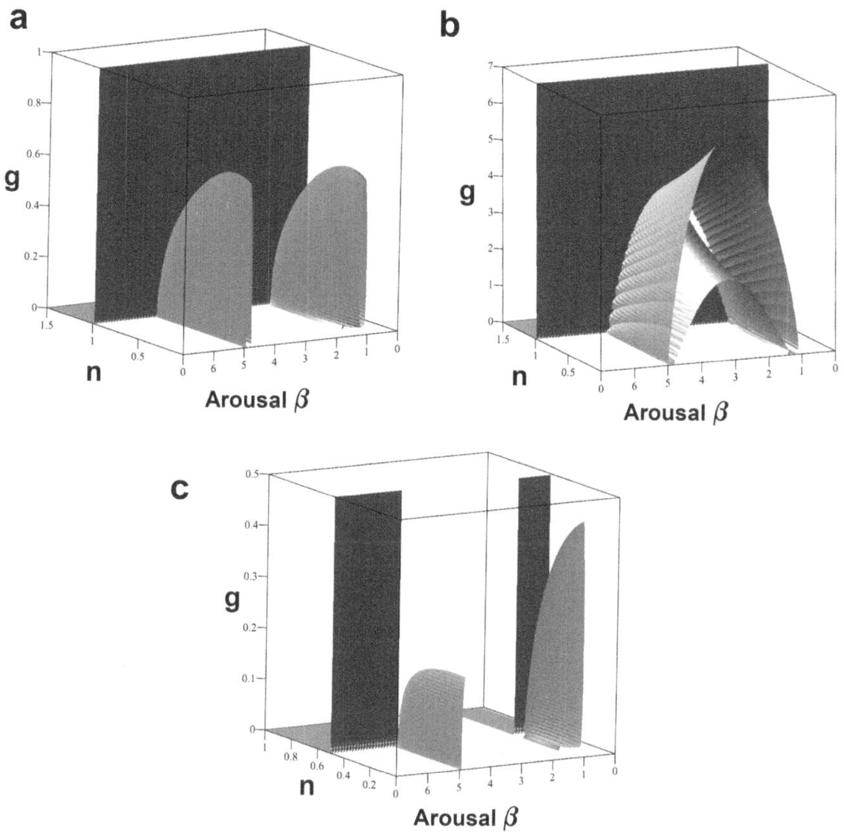

Fig. 3.4 Cognitive temperature $g(n, \beta)$ versus 'arousal' for **a** Cauchy, **b** Nameless $1/(1+x)^2$ and **c** Levy distributions. Only the second, having the simple monotonic-decreasing hazard rate $1/(1+x)$, shows initial classic inverse-U behavior that, nonetheless, collapses for a fully-siloed $n \geq 1$ system

own environment. There is no delay or other dysfunction imposed by the necessity of negotiating and/or evading micromanagement from above. The literature on Mission Command is too large to fully cite here. See (Wallace 2025) or (Matzenbacher 2018). The approach originates from Prussian military doctrine under which units too dispersed to communicate still functioned as virtual mirror-images of each other from extensive previous cross-communication and training (Dupy 1977; Storr 2007). As Fig. 3.3 indicates, the surprising stability of Fig. 3.2 breaks down when the level of command exceeds $k = 1$, even in the absence of 'fat-tail' extreme events.

Such events, of course, precipitate the kinds of daunting complexities explored in Fig. 3.4.

Yes, yes, and truly, we already know much of this, from the distorting and debilitating effects of too many extreme events, to the strong advantages of Workplace Empowerment/Mission Command. What we have done here is to recover something of that knowledge and understanding from a fairly simple set of probability models—a baker's dozen or so equations—built by closely following the asymptotic limit theorems of information and control theories. The next step would be to convert those and such probability models into robust statistical tools—like regression equations, t-tests, and the like—for the analysis of real-time, real-world data.

A skilled reader can take this material and run with it.

3.8 Appendix

The dynamics of the system represented in Fig. 3.2, a simple exponential model $\rho(x) = \exp[-x]$, with varying n in Eq. (3.9), can be explored under stochastic burden by expanding from the direct relation $dZ/dt = \beta - \alpha Z$ to the stochastic differential equation (Cyganowski et al. 2002)

$$dZ_t = (\beta - \alpha Z_t)dt + \sigma Z_t dB_t \tag{3.14}$$

where the second term represents 'ordinary volatility' under Brownian white noise dB_t. The key to the matter is calculating the nonequilibrium steady state condition $\langle dL_t \rangle = 0$ for the cognition rate L from Eq. (3.10), using the Ito Chain Rule. Cyganowski et al. (2002) provide a short MAPLE computer algebra program for such calculations.

We fix the system at the $m = 0$ peak of Fig. 3.2, so that $\alpha = 1,\ \beta = 3$, solve the nss relation $\langle dL_t \rangle = 0$ for σ, and then rotate the graph to give the solution set $\{Z,\ n,\ \sigma\}$ as shown in Fig. 3.5. For $\sigma > \approx 0.78$, no solutions are possible. That is, there is a critical value for stochastic burden, in this model.

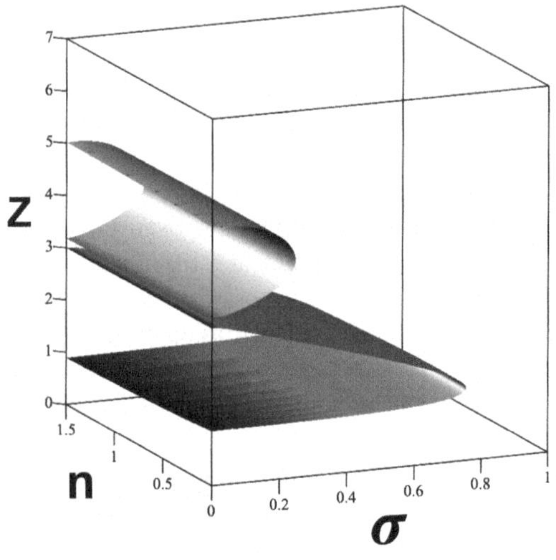

Fig. 3.5 The system is fixed at the $m = 0$ peak of Fig. 3.2, with $\alpha = 1,\ \beta = 3$. Solving the nss relation $\langle dL_t \rangle = 0$ for σ, and rotating the graph, gives the solution set $\{Z,\ n,\ \sigma\}$. For $\sigma > \approx 0.78$, no solutions are possible, imposing a critical value for stochastic burden

References

Atlan, H., Cohen, I.: Immune information, self-organization, and meaning. Int. Immunol. **10**, 711–717 (1998)
Barabasi, A., Albert, R.: Emergence of scaling in random networks. Science **286**, 509–512 (1999)
Bennett, C.H.: The thermodynamics of computation. Int. J. Theor. Phys. **21**, 905–940 (1982)
Brown, R.: From groups to groupoids: a brief survey. Bull. Lond. Math. Soc. **19**, 113–134 (1987)
Cover, T., Thomas, J.: Elements of Information Theory, 2nd edn. Wiley, New York (2006)
Cyganowski, S., Kloeden, P., Ombach, J.: From Elementary Probability to Stochastic Differential Equations with MAPLE. Springer, New York (2002)
de Groot, S., Mazur, P.: Non-Equilibrium Thermodynamics. Dover, New York (1984)
Dembo, A., Zeitouni, O.: Large Deviations and Applications, 2nd edn. Springer, New York (1998)
Derman, E., Miller, M., Park, D.: The Volatility Smile. Wiley, New York (2016)
Dolan, B., Janke, W., Johnston, D., Stathakopoulos, M.: Thin fisher zeros. J. Phys. A **34**, 6211–6223 (2001)
Dretske, F.: The explanatory role of information. Phil. Trans. R. Soc. A **349**, 59–70 (1994)
Dupy, T.: A Genius for War: The German Army and General Staff 1807–1945. Macdonald and Janes, London (1977)
Feynman, R.: Lectures on Computation. Westview Press, New York (2000)
Fisher, M.: Lectures in Theoretical Physics, vol. 7. University of Colorado Press, Boulder (1965)
Genewein, T., Leibfried, F., Grau-Moya, J., Braun, D.: Bounded Rationality, Abstraction, and Hierarchical Decision-making: An Informationtheoretic Optimality Principle. Frontiers in Robotics and AI (2015). https://doi.org/10.3389/frobt.2015.00027
Gerstner, L.: Culture Is Everything, Harvard Business Review (2003). Available online: https://www.alumni.hbs.edu/stories/Pages/story-bulletin.aspx?num=2039
Gigerenzer, G.: Embodied heuristics. Front. Psychol. **12**, 711289 (2021). https://doi.org/10.3389/fpsyg.2021.711289
Harush, U., Barzel, B.: Dynamic patterns of information flow in complex networks. Nat. Commun. **8**, 2181 (2017)
Khinchin, A.: The Mathematical Foundations of Information Theory. Dover, New York (1957)
Laidler, K.: Chemical Kinetics, 3rd edn. Harper and Row, New York (1987)
Landau, L., Lifshitz, E.: Statistical Physics, 3rd Ed. Part 1, Elsevier, New York (2007)
Maturana, H., Varela, F.: Autopoiesis and Cognition: The Realization of the Living. Reidel, Boston (1980)
Matzenbacher, B.: The US army and mission command: philosophy versus practice. Mil. Rev. **2018**(61), 71 (2018)
Mezo I., Keady, G.: Some Physical Applications of Generalized Lambert Functions (2015). arXiv:1505.01555v2 [math.CA] 22 Jun 2015
Mintzberg, H.: The Nature of Managerial Work. Harper and Row, New York, NY, USA (1973)
Mintzberg, H.: The Structure of Organizations, Readings in Strategic Management, pp. 322–352. Macmillan, London, UK (1989)
Nair, G., Fagnani, F., Zampieri, S., Evans, R.: Feedback control under data rate constraints: an overview. Proc. IEEE **95**, 108138 (2007)
Pettini, M.: Geometry and Topology in Hamiltonian Dynamics and Statistical Mechanics. Springer, New York (2007)
Porter, M.: Competitive Strategy. Free Press, New York, NY, USA (1980)
Porter, M.: The five competitive forces that shape strategy. Harv. Bus. Rev. **86**(78), 93 (2008)
Ruelle, D.: Cluster property of the correlation functions of classical gases. Rev. Mod. Phys. 580–584 (1964)
Storr, J.: A command philosophy for the information age: the continuing relevance of mission command. Defense Stud. **119**, 129 (2007)
Taleb, N.N. (ed.).: Statistical Consequences of Fat Tails: Real World Preasymptotics, Epistemology, and Applications. STEM Academic Press (2020)

Wallace, R., Fricchione, G.: Stress-induced failure of embodied cognition: a general model. Biosystems **239**, 105193 (2024)
Wallace, R.: Detailed command versus mission command: a cancer-stage model of institutional decision-making. Stats **8**, 27 (2025). https://doi.org/10.3390/stats8020027
Weinstein, A.: Groupoids: unifying internal and external symmetry. Not. Am. Math. Assoc. **43**, 744–752 (1996)
Zilberstein, S.: Using anytime algorithms in intelligent systems. AI Mag. **17**, 73–83 (1996)

Chapter 4
Moving On

4.1 Introduction

In spite of the apparently dense mathematical overhead of the previous chapters, we have only provided an entry-level treatment. As perusal of the specialist journals dedicated to 'ordinary' statistics indicates, the conversion of the 'ordinary' asymptotic limit theorems of probability theory to useful, validated, and reliable statistical tools is, under the best of circumstances, a fraught enterprise. Turning the asymptotic limit theorems of information and control theories into a robust statistical toolbox can be no less daunting. The statistical physics/nonequilibrium thermodynamics approach taken above as a starting point reflects more the author's particular training than the necessities of the formalism, dimensional analysis notwithstanding. Here, we will examine avenues for extending the theory. Readers can come up with their own.

4.2 Higher Dimensional Systems

We have, thus far, assumed matters are sufficiently characterized by the single scalar parameter Z that mixes material resource/energy supply with internal and external flows of information. Following Wallace (2021)—invoking techniques akin to Principal Component Analysis—it can be argued that there may often be more than one independent composite entity irreducibly driving system dynamics. It may often be necessary to replace the scalar Z with an n-dimensional vector $\mathbf{Z}$ having orthogonal components that, together, account for most of the total variance in the rate of supply of essential resources. The underlying first-order dynamic equations then become considerably more complicated:

© The Author(s), under exclusive license to Springer Nature Switzerland AG 2026
R. Wallace, *Bounded Rationality and Its Discontents*, Studies in Systems, Decision and Control 664, https://doi.org/10.1007/978-3-032-18423-8_4

$$
\begin{aligned}
&\rho(F(\mathbf{Z})/g(\mathbf{Z})) = Q(g(\mathbf{Z})) \\
&S = -F + \mathbf{Z} \cdot \nabla_{\mathbf{Z}} F \\
&\partial \mathbf{Z}/\partial t \approx \hat{\mu} \cdot \nabla_{\mathbf{Z}} S = f(\mathbf{Z}) \\
&\quad -\nabla_{\mathbf{Z}} F + \nabla_{\mathbf{Z}}(\mathbf{Z} \cdot \nabla_{\mathbf{Z}} F) = \\
&\hat{\mu}^{-1} \cdot f(\mathbf{Z}) \equiv f^*(\mathbf{Z}) \\
&\left((\partial^2 F/\partial z_i \partial z_j)\right) \cdot \mathbf{Z} = f^*(\mathbf{Z}) \\
&\left((\partial^2 F/\partial z_i \partial z_j)\right) |_{\mathbf{Z}_{nss}} \cdot \mathbf{Z_{nss}} = \mathbf{0}
\end{aligned} \tag{4.1}
$$

F, f, g, ρ, Q, and S are scalar functions, while $\hat{\mu}$ is an n-dimensional square matrix of diffusion coefficient analogs.

The expression $((\partial F/\partial z_i \partial z_j))$ is an n-dimensional square matrix of second partial derivatives, and $f(\mathbf{Z})$ is a vector function. The last relation imposes a deterministic nonequilibrium steady state condition, i.e. $f^*(\mathbf{Z}_{nss}) = \mathbf{0}$. This may or may not be inherently stable.

Most simply, $Q(g(\mathbf{Z})) \to g(\mathbf{Z})$, with $\mathbf{Z}(t) \to \mathbf{Z}_{nss}$.

This is an overdetermined system of partial differential equations (Spencer 1969) if $n \geq 2$. For a general-form of $f^*(\mathbf{Z})$, the system is inconsistent, resulting in as many as n different expressions for $F(\mathbf{Z})$, and hence the same number of temperature-analog measures.

This conclusion is not surprising. The fifth expression in Eq. (4.1), where $f^*(\mathbf{Z}) \neq \mathbf{0}$, represents a multi-component, cross-interacting, cross-talking, system that, if acting in real time, will almost always be far from any possible steady state condition, equilibrium or nonequilibrium.

Such systems must be inherently unstable, requiring, by the Data Rate Theorem, constant input of control information that will necessarily lag perturbation. Such an enterprise will not be simply characterized by a single cognitive temperature-analog g. Each cognitive component, if the system is far from a steady state, should be expected to have its own g-value, in addition to structure imposed by the multidimensional nature of $\mathbf{Z}$.

The vector function $f(\mathbf{Z})$ becomes the basis for extension of stochastic analysis as a multidimensional stochastic differential equation

$$
d\mathbf{Z}_t = f(\mathbf{Z_t})dt + \rho(\mathbf{Z_t})d\mathbf{B}_t \tag{4.2}
$$

where $d\mathbf{B}_t$ represents n-dimensional Brownian noise.

This development, where the dimension is $n \geq 2$, invokes the dangerous world of Appleby et al. (2008), in which it is almost always possible to find a functional form $\rho(\mathbf{Z})$ that destabilizes an inherently stable function $f(\mathbf{Z})$, or—by contrast—stabilizes an inherently unstable one.

Wallace (2021), further, shows that higher dimensional systems suffer the burdens of Lie Group methods in the study of their deterministic dynamics.

Likewise, the argument can be extended to systems with colored noise, via the Dolean-Dade exponential, contingent on no small algebraic overhead (Protter 2005).

We provide a relatively simple, if somewhat ambiguous, example of stochastic extension, based on the relations of Eq. (3.10):

$$
\begin{aligned}
\rho(C,k,x) &= \frac{C^k x^{k-1} e^{-Cx}}{\Gamma(k)} \\
g &= -\frac{FC}{(n+k)\, W\left(m, -\frac{F\left(F^{k-1}\right)^{-\frac{1}{n+k}} \Gamma(n+k)^{\frac{1}{n+k}}}{n+k}\right)} \\
L &= \frac{\Gamma\left(n+k, \frac{CR_0}{g}\right)}{\Gamma(n+k)}
\end{aligned} \tag{4.3}
$$

We set $C = R_0$ and $k = 1$—giving a Boltzmann Distribution—but allow both Z *and the exponent* n representing degree of siloization to be stochastic variates following the 'exponential' relations

$$
\begin{aligned}
dZ_t &= (3 - Z_t)dt + \sigma Z_t dB_t^1 \\
dn_t &= (1 - n_t)dt + \omega n_t dB_t^2
\end{aligned} \tag{4.4}
$$

Again, the dB_t^k are Brownian white noise and the stochastic terms are taken as 'simple volatility.'

The two-dimensional nonequilibrium steady state condition $\langle dL_t \rangle = 0$ then becomes very complicated indeed under the Ito Chain Rule. We define $F(\alpha, \beta, C_1, C_2, Z)$ as in Eq. (3.6), with, from Eq. (4.4) $\alpha = 1,\ \beta = 3$. Here, again, $C_1 = -1,\ C_2 = 3$. In the expression for L, n is assigned the values $0,\ 1/2,\ 1,\ 2$. Using the *implicitplot3d* numerical procedure of the computer algebra program MAPLE 2025, we calculate $Z(\omega, \sigma)$ as in Fig. 4.1. Rising 'noise' as indexed by ω and σ, constrains possible solutions.

4.3 Higher Order Arguments

Further extensions of theory are possible, if not entirely straightforward, for example, expanding both the very idea of 'entropy' and the Onsager gradient model defining system dynamics, to higher orders. This can be done via formal power series algebraic ring structures, following the lead of Jackson et al. (2017), who see the Legendre transform as mapping the coefficients of one formal power series into the coefficients of another formal power series, modulo dimensional restrictions and consistency. One implication of this approach is that the 'spectra' of the formal power series coefficients characterize essential features of the systems under study, and will likely display particular dynamics under changing patterns of selection pressure and affordance.

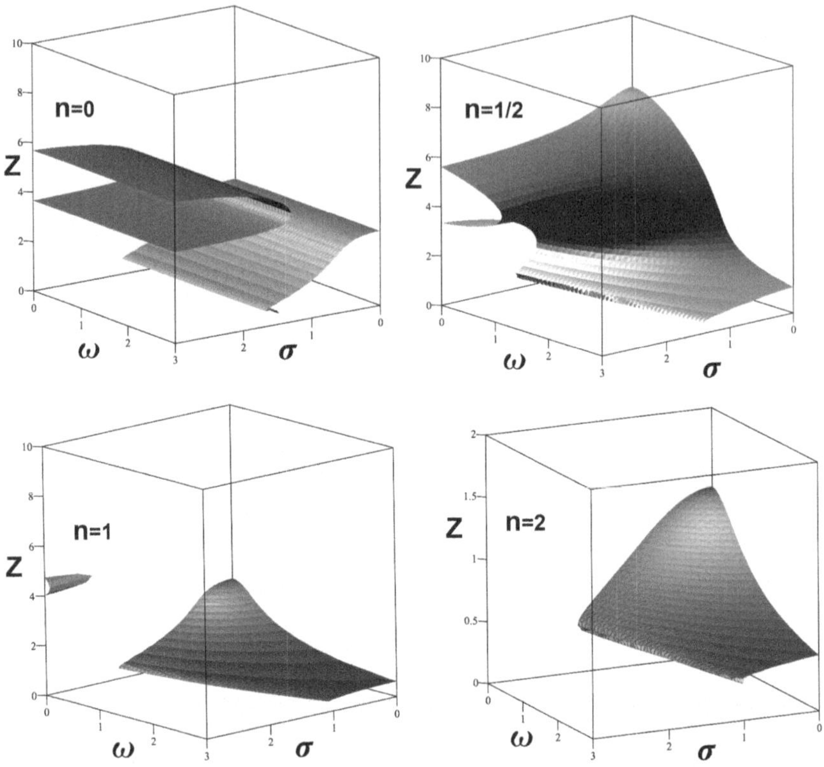

Fig. 4.1 Solution sets for $\langle dL_t \rangle = 0$ from Eq. (4.3) under the SDE of Eq. (4.4) for siloing index $n = 0,\ 1/2,\ 1,\ 2$. Rising 'noise' ω and σ constrain possible solutions. Under the conditions of Eq. (4.4), $n = 2$ is problematic. Indeed, given two or more stochastic variables, the considerations of Appleby et al. (2008) become manifest, with much depending on the exact forms of the 'variance' functions

A 'simple' generalization—omitting crossterms—might take the form

$$
S = -F + Z dF/dZ + \sum_{k=2}^{n} \varepsilon_k Z^k d^k F/dZ^k
$$

$$
dZ/dt = \sum_{k=1}^{m} \mu_k Z^{k-1} d^k S/dZ^k \qquad (4.5)
$$

Here, we use

$$
\begin{aligned}
&S = -F + ZdF/dZ + \varepsilon Z^2 d^2F/dZ^2 \\
&dZ/dt = f(Z) = \beta - \alpha Z = dS/dZ + \mu Z d^2 S/dZ^2 \\
&F = \frac{\beta\,(Z\ln(Z) - Z)}{\varepsilon + 1} - \frac{\alpha\, Z^2}{4\varepsilon\mu + 4\varepsilon + 2\mu + 2} - \frac{\mu^2 C_1 Z}{Z^{\frac{1}{\mu}}\,(\mu - 1)} \\
&\quad + \frac{\varepsilon^2 C_2}{(\varepsilon + 1)\, Z^{\frac{1}{\varepsilon}}} + C_3 Z + C_4
\end{aligned}
\tag{4.6}
$$

We take $\alpha = 1$, again assign β as the index of 'arousal', setting $\mu = \varepsilon = 1/2$, $C_1 = C_2 = 1$, $C_3 = -1$, $C_4 = 3$.

We first study the familiar Boltzmann Distribution example from above, i.e., $\exp[-F/g] = g$. Recall that this gives

$$
\begin{aligned}
g &= \frac{-F}{W(m,\ -F)} \\
L &= \exp[-1/g]
\end{aligned}
\tag{4.7}
$$

where, again, $W(m, x)$ is the Lambert W-function of order m. Taking $m = 0$ and F as above we find the graph of cognition rate versus arousal to be the familiar Yerkes-Dodson inverse-U.

Next, we reprise the more complex calculations of Eqs.(3.10) and (4.3) that lead to Fig. 3.2, again setting $k = C = R_0 = 1$ but now using F from Eq. (4.6) as in Fig. 4.2. The result is similar to, but subtly different from, Fig. 3.2.

The orders of the model do indeed count, but this must be an empirical matter, much as the assignment of the order and form of a regression equation (Fig. 4.3).

Extending Eq. (4.5) via crossterms in a vectorized **Z** is left as an exercise.

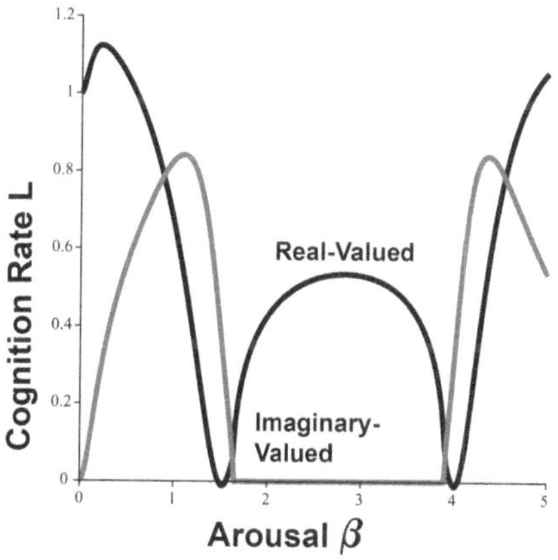

Fig. 4.2 The simplest extension of Eq. (4.5) still generates the familiar Yerkes-Dodson inverse-U of cognition rate versus arousal

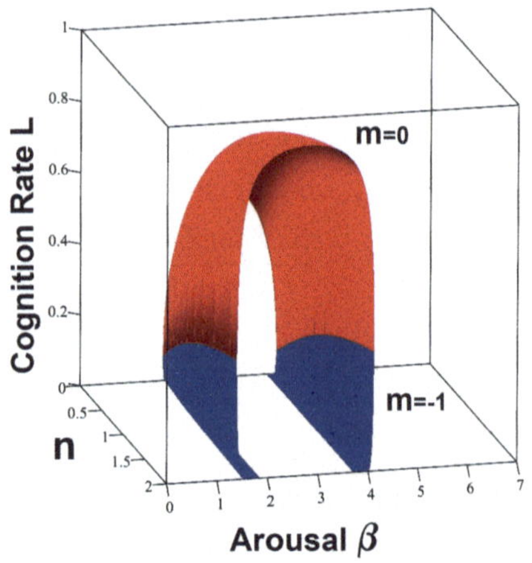

Fig. 4.3 Recalculation of Fig. 3.2 using the F relation of Eq. (4.6). The Yerkes-Dodson inverse-U reemerges, but in subtly different form

4.4 Modes of Friction

Another 'obvious' model extension involves different modalities of 'friction', i.e., changes to the form of $f(\mathbf{Z})$ in the third expression of Eq. (4.1), Eq. (4.2), and the second expression of Eq. (4.6). That is, moving beyond the exponential model $dZ/dt = \beta - \alpha Z$. In particular we will reconsider the Boltzmann model of Eq. (4.7) for expressions $Z(t)$ having the forms

$$
\begin{aligned}
&Z(t) = \beta \exp[-\alpha/t], \;\; f(Z) = \frac{Z}{\alpha}\,(\ln(Z/\beta))^2 \\
&Z(t) = \frac{\beta}{\alpha}\,(1 - \exp[-\alpha\, t])\,, \;\; f(Z) = \beta - \alpha Z \\
&Z(t) = \sqrt{\beta/\alpha}\,\tanh(\sqrt{\beta\alpha}\, t), \;\; f(Z) = \beta - \alpha Z^2
\end{aligned}
\tag{4.8}
$$

These are the Arrhenius, exponential, and Tanh models.

Imposing the first-order version of Eq. (4.6),

$$
\begin{aligned}
&F = \frac{\ln\left(\frac{Z}{\beta}\right)^2 Z^2}{2\alpha} - \frac{3Z^2 \ln\left(\frac{Z}{\beta}\right)}{2\alpha} + \frac{7Z^2}{4\alpha} + C_1 Z + C_2 \\
&F = \beta\,(Z\ln(Z) - Z) - \frac{Z^2\alpha}{2} + C_1 Z + C_2 \\
&F = -\frac{\alpha\, Z^3}{6} + \beta\,(Z\ln(Z) - Z) + C_1 Z + C_2
\end{aligned}
\tag{4.9}
$$

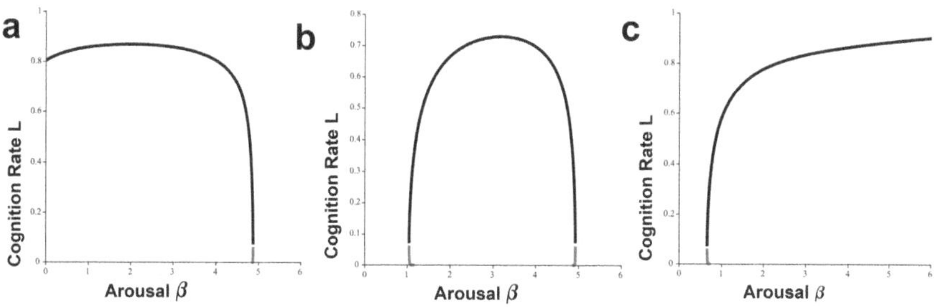

Fig. 4.4 Best-case cognition rate versus arousal for **a** Arrhenius, **b** exponential, and **c** tanh friction models. Respectively, $C_1 = C_2 = -7$, $C_1 = -1$, $C_2 = 3$, and $C_1 = -3$, $C_2 = 3$. Friction form counts

Figure 4.4 displays, for the model of Eq. (4.7), the 'most characteristic' forms of cognition rate versus arousal β for the $m = 0$ and $m = -1$ branches of the Lambert W-function, given the boundary conditions as indicated in the figure caption.

The nature of the friction counts.

4.5 An Alternate Formalism

We have, thus far, viewed the world through the lens of Rate Distortion Control Theory, via Fig. 1.1. It is possible to reinterpret matters from the perspectives of information source uncertainty rather than rate distortion using an avalanche of arguments:

- Recalling Atlan and Cohen (1998), cognition requires choice that reduces uncertainty. Such reduction in uncertainty directly implies the existence of an information source 'dual' to that cognition at each scale and level of organization.
- Cognitive processes are usually inherently unstable, requiring control information at a rate greater than the process generates its own topological information. That is, a cognitive information source X is generally paired with a regulatory information source $\widehat{X}$.
- Environments impose temporal event sequences of very high probability: night follows day, hot seasons follow cold, wet season follows dry, and so on. Thus environments impose their own 'meaningful statements' onto entities and interactions embedded within them via an information source V.
- 'Large deviations', following Champagnat et al. (2006) and Dembo and Zeitouni (1998), also involve sets of high probability developmental pathways, often governed by 'entropy'-like laws that imply the existence of yet another information source, L_D.

The full system must then be characterized by a joint—individually path dependent—nonergodic information source having source uncertainty

$$H(X, \widehat{X}, V, L_D) \tag{4.10}$$

H is to be characterized jointly by the information sources X, $\widehat{X}$, V and L_D.

Individual dynamic paths can then be assigned a value of that joint source uncertainty, denoted by $H(x)$ for a path x.

This new representation is defined on individual dynamic path values of source uncertainty and is not an 'entropy-like' function defined for all high-probability paths by an underlying probability distribution (Khinchin 1957). Each path has it's own H-value, but the functional form of that value is not known in terms of a single underlying probability distribution across all paths.

To reiterate, the pairing $\{X, \widehat{X}\}$ includes the internal interactive cognitive dual information sources of the system of interest and the associated regulator, V is the information source of the embedding environment that may include the actions and intents of adversaries/symbionts, as well as 'weather'. L_D is the information source of the associated large deviations possible to the system.

Again, we project the matrix of essential resources and their interactions onto a scalar rate index Z.

Following from the properties of nonergodic information sources, equivalence classes of developmental-behavioral-dynamic system paths can be used to define groupoid symmetries, defined fully in terms of the magnitude of individual path source uncertainties *of individual dynamic paths* $H(x_j)$ such that $x_j = \{x_j^0, x_j^1, \ldots x_j^n, \ldots\}$ at times $m = 0, 1, 2, \ldots n \rightarrow \infty$ alone rather than in terms of scalar distortion measures required by the RDT (Khinchin 1957). Individual paths of sufficient length have associated source uncertainty scalar values that are not, however, calculated as standard Shannon 'entropies' across some probability distribution.

One might view the equivalence classes of behavioral/developmental paths as defined by the 'game' the entity of interest is playing at a given time. Paths within each 'game' are taken as equivalent. This division of developmental/behavioral paths defines a groupoid, and phase transition dynamics will be driven by 'groupoid symmetry-breaking'.

It is now possible to replace R in Eqs. (1.2–1.4) and Eqs. (1.12–1.13) with H, and redevelop the subsequent dynamic arguments.

4.6 Symmetry and Symmetry-Breaking

Further development enters the cutting-edge world of group—and groupoid—symmetry and symmetry-breaking for information sources.

The Centre for Theoretical Cosmology at Cambridge (CTC 2021) describes symmetry-breaking in physics this way:

> The basic premise of grand unification is that the known symmetries of the elementary particles resulted from a larger (and so far unknown) symmetry group **G**. Whenever a phase transition occurs, part of this symmetry is lost, so the symmetry group changes. This can be represented mathematically as

$$\mathbf{G} \rightarrow \mathbf{H} \rightarrow \ldots \rightarrow \mathbf{SU_3} \times \mathbf{SU_2} \times \mathbf{U_1} \rightarrow \mathbf{SU_3} \times \mathbf{U_1}$$

> Here, each arrow represents a symmetry breaking phase transition where matter changes form and the groups—**G**, **H**, $\mathbf{SU_3}$, etc.—represent the different types of matter, specifically the symmetries that the matter exhibits, and they are associated with the different fundamental forces of nature.
>
> The cosmological significance of symmetry breaking is due to the fact that symmetries are restored at high temperature (just as it is for liquid water when ice melts). For extremely high temperatures in the early Universe, we will even achieve a grand unified state **G**. Viewed from the moment of creation forward, the Universe will pass through a succession of phase transitions at which the strong nuclear force will become differentiated and then the weak nuclear force and electromagnetism...

In more general terms, Stewart (2017) puts the matter as follows:

> Spontaneous symmetry-breaking is a common mechanism for pattern formation in many areas of science. It occurs in a symmetric dynamical system when a solution of the equations has a smaller symmetry group than the equations themselves... This typically happens when a fully symmetric solution becomes unstable and branches with less symmetry bifurcate.

Symmetries and information theory conjoin in the work of Yeung (2008) who found fundamental relations between information theory inequalities and the theory of finite groups that are similar to those between synchronicity in networks and their permutation symmetries explored by Golubitsky and Stewart (2006). Suppose we have random variables X_1 and X_2 with Shannon uncertainties $H(X_1)$ and $H(X_2)$. The information theory chain rule (Cover and Thomas 2006) finds, for the joint uncertainty $H(X_1, X_2)$, that

$$H(X_1) + H(X_2) \geq H(X_1, X_2) \tag{4.11}$$

This expression has a direct mapping to finite group theory: let G be a finite group, and G_1, G_2 are subgroups of G. Let $\|G\|$ be the order of a group—the number of its elements. Then the intersection $G_1 \cap G_2$ is also a subgroup, and a 'group inequality' can be derived that is the exact analog of Eq. (4.11):

$$\log[\frac{\|G\|}{\|G_1\|}] + \log[\frac{\|G\|}{\|G_2\|}] \geq \log[\frac{\|G\|}{\|G_1 \cap G_2\|}] \tag{4.12}$$

Yeung (2008) defines a probability for a pseudo-random variate associated with a group G as $Pr\{X = a\} = 1/\|G\|$, permitting construction of a group-characterized information source, given that, in general, the joint uncertainty of a set of random variables is not necessarily the logarithm of a rational number. Yeung (2008) shows one-to-one correspondence between unconstrained information inequalities—generalizations of Eq. (4.11)—and finite group inequalities: unconstrained inequalities can be proved by techniques in group theory, and many group-theoretic inequalities can be proven by techniques of information theory. Yeung applies an otherwise obscure unconstrained information inequality to derive, in his Eq. (16.116), a complex and equally obscure—but previously unknown—group inequality.

Extension of the Yeung arguments follows from the central argument of algebraic topology (Hatcher 2001): that it is possible to form algebraic images of topological spaces. The most basic such image is the fundamental group, leading to Van Kampen's Theorem permitting computation of the fundamental group of spaces that can be decomposed into simpler spaces whose fundamental group is already known. Brown et al. (2011) extend these considerations to groupoids that might be applied to more complicated networks. We summarize material on groupoids in the Chapter Appendix.

Yeung's results suggest information theory-based 'cognitive' generalizations that may extend to essential dynamics of cognition, and to the role of directed homotopy and groupoid symmetry-breaking in those dynamics. Information sources inherently involve directed homotopy via the improbability of 'microreversibility', e.g., in English the sequence of symbols 'the' is far more probable than 'eht'.

4.7 Discussion

We have briefly outlined possible extensions of the probability models studied here to 'higher order' relations, much as one might move from a regression model like $Y = mX + b$ to $Y = mX^n + b$, and so on. Unlike the various 'XXX Principle' models that claim to be General Relativity-type statements about the fundamental structure of the cognitive universe, we are more modest, seeking to derive useful statistical tools based on the asymptotic limit theorems of information and control theories. In this we follow a well-trodden track that has used the Central Limit Theorem and Friends to derive the plethora of statistical tools that grease the wheels of the experimental and observational sciences. Statistical tools provide useful benchmarks against which to compare and contrast patterns found in data.

One hopes that the tools developed in and from this work will prove useful to others in their searches for as-yet-to-be-discovered Great Principles engaging with and emerging from cognitive phenomena at and across the many levels and grand scales of the living state.

4.8 Mathematical Appendix: Groupoids

Following Weinstein (1996), a groupoid, G is defined by a base set A upon which some mapping—a morphism—can be defined. Note that not all possible pairs of states (a_j, a_k) in the base set A can be connected by such a morphism. Those that can define the groupoid element, a morphism $g = (a_j, a_k)$ having the natural inverse $g^{-1} = (a_k, a_j)$. Given such a pairing, it is possible to define 'natural' end-point maps $\alpha(g) = a_j, \beta(g) = a_k$ from the set of morphisms G into A, and an associative product in the groupoid $g_1 g_2$ provided $\alpha(g_1 g_2) = \alpha(g_1), \beta(g_1 g_2) = \beta(g_2)$, and $\beta(g_1) = \alpha(g_2)$. Then the product is defined, and associative, $(g_1 g_2) g_3 = g_1 (g_2 g_3)$.

In addition, there are natural left and right identity elements λ_g, ρ_g such that $\lambda_g g = g = g\rho_g$ (Weinstein 1996).

An orbit of the groupoid G over A is an equivalence class for the relation $a_j \sim Ga_k$ if and only if there is a groupoid element g with $\alpha(g) = a_j$ and $\beta(g) = a_k$. Following Cannas da Silva and Weinstein (1999), we note that a groupoid is called transitive if it has just one orbit. The transitive groupoids are the building blocks of groupoids in that there is a natural decomposition of the base space of a general groupoid into orbits. Over each orbit there is a transitive groupoid, and the disjoint union of these transitive groupoids is the original groupoid. Conversely, the disjoint union of groupoids is itself a groupoid.

The isotropy group of $a \in X$ consists of those g in G with $\alpha(g) = a = \beta(g)$. These groups prove fundamental to classifying groupoids.

If G is any groupoid over A, the map $(\alpha, \beta) : G \to A \times A$ is a morphism from G to the pair groupoid of A. The image of (α, β) is the orbit equivalence relation $\sim G$, and the functional kernel is the union of the isotropy groups. If $f : X \to Y$ is a function, then the kernel of f, $ker(f) = [(x_1, x_2) \in X \times X : f(x_1) = f(x_2)]$ defines an equivalence relation.

Groupoids may have additional structure. As Weinstein (1996) explains, a groupoid G is a topological groupoid over a base space X if G and X are topological spaces and α, β and multiplication are continuous maps. A criticism sometimes applied to groupoid theory is that their classification up to isomorphism is nothing other than the classification of equivalence relations via the orbit equivalence relation and groups via the isotropy groups. The imposition of a compatible topological structure produces a nontrivial interaction between the two structures. For example, introducing a metric structure on manifolds of related information sources produces such interaction.

In essence, a groupoid is a category in which all morphisms have an inverse, here defined in terms of connection to a base point by a meaningful path of an information source dual to a cognitive process.

Again, from Weinstein (1996), the morphism (α, β) suggests another way of looking at groupoids. A groupoid over A identifies not only which elements of A are equivalent to one another (isomorphic), but *it also parameterizes the different ways (isomorphisms) in which two elements can be equivalent*, i.e., in our context, all possible information sources dual to some cognitive process. Given the information theoretic characterization of cognition presented above, this produces a full modular cognitive network in a highly natural manner.

Brown (1987) describes the fundamental structure in these terms:

> A groupoid should be thought of as a group with many objects, or with many identities... A groupoid with one object is essentially just a group. So the notion of groupoid is an extension of that of groups. It gives an additional convenience, flexibility and range of applications...
>
> EXAMPLE 1. A disjoint union [of groups] $G = \cup_\lambda G_\lambda$, $\lambda \in \Lambda$, is a groupoid: the product ab is defined if and only if a, b belong to the same G_λ, and ab is then just the product in the group G_λ. There is an identity 1_λ for each $\lambda \in \Lambda$. The maps α, β coincide and map G_λ to λ, $\lambda \in \Lambda$.

EXAMPLE 2. An equivalence relation R on [a set] X becomes a groupoid with $\alpha, \beta : R \rightarrow X$ the two projections, and product $(x, y)(y, z) = (x, z)$ whenever $(x, y), (y, z) \in R$. There is an identity, namely (x, x), for each $x \in X$...

Weinstein (1996) makes an important point:

Almost every interesting equivalence relation on a space B arises in a natural way as the orbit equivalence relation of some groupoid G over B. Instead of dealing directly with the orbit space B/G as an object in the category S_{map} of sets and mappings, one should consider instead the groupoid G itself as an object in the category G_{htp} of groupoids and homotopy classes of morphisms.

The groupoid approach has become quite popular in the study of networks of coupled dynamical systems which can be defined by differential equation models, (e.g., Golubitsky and Stewart 2006).

References

Appleby, J., Mao, X., Rodkina, A.: Stabilization and destabilization of nonlinear differential equations by noise. IEEE Trans. Autom. Control **53**, 68–69 (2008)

Atlan, H., Cohen, I.: Immune information, self-organization, and meaning. Int. Immunol. **10**, 711–717 (1998)

Bluman, G., Kumei, S.: Symmetries and differential equations. Springer, New York (1989)

Brown, R.: From groups to groupoids: a brief survey. Bull. Lond. Math. Soc. **19**, 113134 (1987)

Brown, R., Higgins, P., Sivera, R.: Nonabelian Algebraic Topology: Filtered Spaces, Crossed Complexes, Cubical Homotopy Groupoids, EMS Tracts in Mathematics, vol. 15 (2011)

Champagnat, N., Ferriere, R., Meleard, S.: Unifying evolutionary dynamics: from individual stochastic process to macroscopic models. Theor. Popul. Biol. **69**, 297–321 (2006)

CTC.: (2021) http://www.ctc.cam.ac.uk/outreach/origins/cosmicstructuresone.php

Dembo, A., Zeitouni, O.: Large Deviations and Applications, 2nd edn. Springer, New York (1998)

Golubitsky, M., Stewart, I.: Nonlinear dynamics and networks: the groupoid formalism. Bull. Am. Math. Soc. **43**, 305364 (2006)

Hatcher, A.: Algebraic Topology. Cambridge University Press, New York (2001)

Jackson, D., Kempf, A., Morales, A.: A robust generalization of the Legendre transform for QFT. J. Phys. A **50**, 225201 (2017)

Khinchin, A.: The Mathematical Foundations of Information Theory. Dover, New York (1957)

Protter, P.: Stochastic Integration and Differential Equations: A New Approach, 2nd edn. Springer, New York (2005)

Spencer, D.: Overdetermined systems of linear partial differential equations. Bull. Am. Math. Soc. **75**, 179–239 (1969)

Stephani, H., MacCallum, M.: Differential Equations: Their Solutions Using Symmetries. Cambridge University Press, New York (1989)

Stewart, I.: Spontaneous symmetry-breaking in a network model for quadruped locomotion. Int. J. Bifurcat. Chaos **14**, 1730049 (2017) (online)

Wallace, R.: Carl von Clausewitz, the Fog-of-War, and the AI Revolution: The Real World is Not a Game of Go. Springer, New York (2018)

Wallace, R.: Cognitive Dynamics on Clausewitz Landscapes: The Control and Directed Evolution of Organized Conflict. Springer, New York (2020)

Wallace, R.: How AI founders on adversarial landscapes of fog and friction. J. Defense Model. Simul. (2021). https://doi.org/10.1177/1548512920962227

Weinstein, A.: Groupoids: unifying internal and external symmetry. Not. Am. Math. Assoc. **43**, 744752 (1996)
Yeung, H.: Information Theory and Network Coding. Springer, New York (2008)

Chapter 5
Psychopathology Isn't a Bug. It's a Feature

5.1 Introduction

Wang et al. (2025), Dulberg et al. (2023), and others, argue that AI must evolve into a distributed ecosystem where multiple AI entities interact, specialize, and collectively enhance intelligence. A central assertion is that a modular architecture can learn more efficiently and effectively to manage multiple homeostatic objectives, in nonstationary environments, and as the number of objectives increases.

The origin of this approach can be found in the work of Wolpert and Macready (1995, 1997)—more recently, Shalev-Schwartz and Ben-David (2014)—demonstrating that an optimizer tuned to be 'best' on one particular kind of problem must inevitably be 'worst' at some complementary problem set. Complex problems that divide 'naturally' into modular subcomponents can then be—it is hoped—efficiently addressed by an interacting system of cognitive modules each individually tuned to be 'best' on the different problem components most representative of the embedding challenge environment. Success in such a 'game' requires matching cognitive system subcomponents to problem subcomponents in real time. This is not easy. As an anonymous commentator put the matter (2025),

> Apparently a proved theorem cannot be entirely circumvented by a framework consistent in the same formal foundation system, so [Mixture of Experts] just redistributes the challenge across the framework's underlying components.

In particular, the maintenance of dynamic homeostasis under the Clauswitzian 'selection pressures' of fog, friction, and adversarial intent is far from trivial, and, among humans, even after some considerable evolutionary pruning, high-level cognitive process remains liable to debilitating 'culture-bound syndromes' often associated with environmental stress and related, inherently toxic, exposures and subsequent developmental trajectories (e.g., Wallace 2017, Ch. 3).

© The Author(s), under exclusive license to Springer Nature Switzerland AG 2026

R. Wallace, *Bounded Rationality and Its Discontents*, Studies in Systems, Decision and Control 664, https://doi.org/10.1007/978-3-032-18423-8_5

We are, then, very broadly concerned with maintenance of cognitive stability in a modular system under the burdens of what in control theory is called 'topological information' (Nair et al. 2007) imposed by disruptive circumstance, internally or externally generated.

Maturana and Varela (1980) hold that the living state is cognitive at every scale and level of organization. Atlan and Cohen (1998) characterize cognition as the ability to choose an action—or a proper subset—from the full set of those available, in response to internal and/or environmental signals. Such choice reduces uncertainty, in a formal manner, and implies the existence of an information source dual to the cognitive process studied. The argument is direct, and places cognition under the constraints of the asymptotic limit theorems of information theory (Cover and Thomas 2006; Khinchin 1957; Wallace 2022).

More recently, the Data Rate Theorem (Nair et al. 2007) sets the conditions under which an inherently unstable dynamic process can be stabilized by the imposition of control information at a rate greater than that process generates its own 'topological information'. The model is of a vehicle driven at high speed along a twisting, potholed roadway. The driver must impose steering, braking, and gear-shift information at a rate greater than the road imposes its own topology, given the chosen speed.

Most real-world cognitive phenomena, from the living state to such embodied machine cognition as High Frequency Stock Trading (Dreyfus 1972; Fjelland 2020; Wallace 2024), must act in concert with—be actively paired with—a regulator to successfully meet the demands of real-world, real-time function:

- Blood pressure must respond to changing physiological need, but be kept within tolerable limits.
- The immune system must engage in routine cellular maintenance and pathogen control, without attacking healthy tissue via autoimmune disease.
- The stream of consciousness in higher animals must be kept within 'riverbanks' useful to the animal and its social groupings, particularly among organized hominids, the most deadly predators on Earth for the best part of a half-million years.
- Human institutions and organizations, via 'rule-of-law' or 'doctrine' constraints, are notorious for cognition-regulation pairing.

Here, using the asymptotic limit theorems of information and control theories, we will explore, in a sense, the 'metabolic', 'free energy' and other such costs of this necessary pairing as the number of basic, interacting submodules rises, uncovering a truly remarkable set of scaling properties across the 'fundamental underlying probability distributions' likely to power and constrain a variety of important cognitive phenomena.

The probability models arising 'naturally' from this effort can serve as the basis of new statistical tools for the analysis of data related to the stability of cognitive phenomena across a broad range of phenomena. However, like the regression models derived from the Central Limit Theorem and friends, such tools do not, of themselves, 'do the science'. This work has, consequently, focused on a range of models and modeling strategies. Not everything follows a $y = mx + b$ pattern. Sometimes things go as $y = mx^2 + b$ or worse, hence our insistence on the analysis of dynamics across different underlying probability distributions.

Some current attempts at comprehensive cognitive models, e.g., 'integrated information theory', 'the free energy principal', 'quantum consciousness', have made claims of existential universality akin to a general relativity of biology and mind. This work attempts, far more modestly, to derive tools that can aid in the analysis of observational and experimental data, the only sources of new knowledge as opposed to new speculation.

That being said, that being said, we do not at all denigrate the model-based speculations that might arise from our approach, provided they, ultimately, are constrained by effective feedback from data and experiment.

The development, unfortunately, is far from straightforward, and requires the introduction of some considerable methodological boilerplate, extending relatively simple 'first order' approaches abducted from statistical physics and the Onsager model of nonequilibrium thermodynamics, generalizing, in a sense, the results of Khaluf et al. (2017), Kringlebach et al. (2024) and other recent work.

An often-overlooked but essential matter is that information sources are not, like physical processes are assumed to be, microreversible, i.e., in English 'the' has a much higher probability than the sequence 'eht', implying a directed homotopy ultimately leading to groupoid symmetry-breaking cognitive phase transitions. Much—but not all—of this will be safely hidden in the relatively simple first-order formalism developed below.

Another characteristic misunderstanding is the interpretation of Shannon uncertainty, from its mathematical expression, as an actual entropy. Following Feynman (1996) and Bennett (1982), information should be taken as a form of free energy that can be used to build partition functions leading to higher-order free energy constructs. This is, as Feynman remarks, a subtle matter.

More generally, as Khinchin (1957) indicates, for nonergodic stationary systems, each message stream converges to its own source uncertainty that cannot be expressed in 'entropy' format. A standard approach from statistical physics, however, still allows construction of a partition function in such cases, hence an iterated free energy, and, via a Legendre Transform, a compound entropy measure. The gradient of that compound entropy can be taken as a thermodynamic force, defining the rate-of-change of underlying variates, permitting standard stochastic extensions of the models. Again, however, this provides only an entry-level, first-order attack on a barbed-wire thicket of cognitive phenomena and their inevitable pathologies.

Although modern medicine can often perform seeming miracles against an increasingly wide spectrum of infectious and chronic diseases, the culture-bound syndromes of 'madness' remain largely intractable.

The narrowly-trained designers and builders of cutting-edge artificial intelligence systems and entities are entering a Brave New World.

5.2 An Introduction to Method

Resources

Real-world, real-time systems are necessarily embedded in-and-of an environment that includes themselves. There must be at least three resource streams necessary for correct function:

- The rate at which elements of the system can communicate with each other, characterized by some channel capacity $\mathscr{C}$.
- The rate at which 'sensory information' is available from the embedding circumstance, according to a channel capacity $\mathscr{H}$.
- The rate at which 'material/materiel' resources can be provided, a rate $\mathscr{M}$. For an organism, this might be measured by the rate at which metabolic free energy is provided.

These rates cross-correlate and interact, generating a (minimally) 3×3 matrix $\mathbf{Z}$. An $n \times n$ matrix will have n scalar measures r_i invariant under similarity transformations that are determined by the standard expression

$$p(\gamma) = \det[\mathbf{Z} - \gamma \mathbf{I}] = \sum_{k=0}^{n} (-1)^k r_k \gamma^{n-k} \tag{5.1}$$

$\mathbf{I}$ is the identity matrix, 'det' the determinant, and γ a real-valued parameter. The first is the matrix trace, and the last the determinant. We build a scalar resource rate measure from the r_k.

An elementary index might be $Z = \mathscr{C} \times \mathscr{H} \times \mathscr{M}$, but matters are not likely quite that simple.

We do, however, impose a first-order model, assuming that a single scalar index $Z(r_1, ..., r_n)$ can be constructed that sufficiently represents information and material resource rates. This argument is analogous to principal component analysis based on a correlation matrix. Wallace (2021) outlines difficulties inherent to more than a single such index.

Rate Distortion Control Theory

An 'embodied' formalism emerges via a 'Rate Distortion Control Theory' based on the Rate Distortion Theorem (RDT) (Cover and Thomas 2006; Khinchin 1957). The RDT addresses transmission under conditions of noise over an information channel. There will be, for that channel and a particular scalar measure of average distortion D between what is sent and what is received, a minimum necessary channel capacity $R(D)$.

The theorem asks, in effect, what is the 'best' channel for transmission of a message with the least possible average distortion. $R(D)$ can be defined for nonergodic information sources via a limit argument based on the ergodic decomposition of a nonergodic source into a 'sum' of ergodic sources. The RDT can be reconfigured in a control theory context, if we envision a system's topological information from the Data Rate Theorem as another form of noise, adding to the average distortion D. See Fig. 5.1.

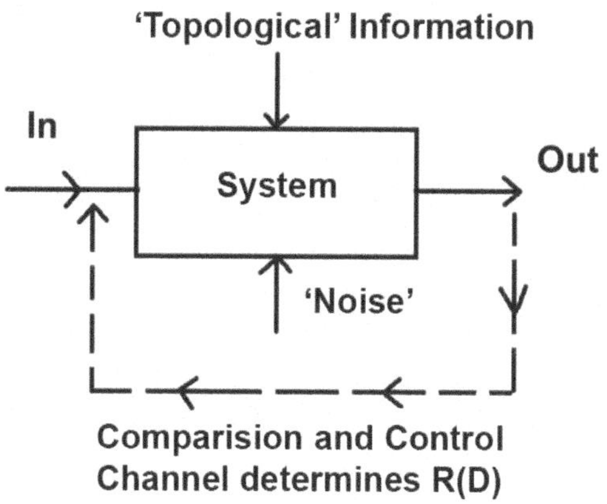

Fig. 5.1 The Rate Distortion Control Theory model. An inherently unstable system's 'topological information', according to the Data Rate Theorem (Nair et al. 2007), is added to the effects of 'noise' to determine the minimum channel capacity $R(D)$ needed to assure transmission with average distortion D. The punctuation implied by the DRT emerges here if there is a critical maximum average distortion beyond which control fails

The punctuation implied by the Data Rate Theorem (DRT) (Nair et al. 2007) happens if there is a critical maximum average distortion beyond which control fails.

Embodied cognition, characterized by continuous interaction with a real world having its own intents and capabilities—as indexed by the distortion between what is wanted and what is observed—is inherently a phenomenon of 'bounded rationality' in the sense of Ortega and Braun (2013). Cognition without such embodied interaction, however, as Dreyfus (1972), Fjelland (2020), and many others have noted, risks a narrow boundedness without rationality.

The 'Tuning Theorem'

Parallel to both Rate Distortion Control Theory and the No Free Lunch Theorem is the 'tuning theorem' inverse of the Shannon Coding Theorem (SCT). The SCT holds that an 'infinite' message stream encoded so that its probabilities are 'typical' with respect to the channel along which it is transmitted can be sent with arbitrarily small error so long as the transmission rate is $\leq C$, the 'capacity' of that channel.

A relatively simple argument (Wallace 2017, Sect. 12.2) shows that, conversely, if a message must be sent at a maximum possible rate, there is a channel which is effectively 'transmitted by the message' at an effective dual channel capacity C^*, where the channel probabilities are now made 'typical' with respect to that particular message. A given inherently unstable system's optimal 'control message' can thus be transmitted at a maximum rate by a 'tuned channel' that is optimal for that message. Different optimal control messages must be transmitted by different optimal channels, hence the No Free Lunch Theorem: there is no single optimal channel for all possible control messages.

5.3 A Single Cognition-Regulation Pairing

Here, we are interested in the dynamics of interaction between two entities—a 'system-of-interest' and its paired regulating apparatus—engaged separately in 'real-world' evaluations according to Fig. 5.1, where each entity is concerned with its own channel capacity $R_i(D_i)$—increasing R to lower the distortion between what it intends and what it observes.

In rough consonance with the approaches of Ortega and Braun (2013) and Kringlebach et al. (2024), we consider a 'simple' probability distribution underlying overall system dynamics, say $\rho(x)$, and construct an iterated probability for each component of the interaction individually, having the form

$$P_j = \frac{\rho(R_j/g_j(Z_j))}{\rho(R_1/g_1(Z_1)) + \rho(R_2/g_2(Z_2))} \tag{5.2}$$

where $g(Z)$ represents a *temperature analog* that must be further characterized, and each R_j, following Feynman's (1996, p.146) adaptation of Bennett's (1982) analysis, is itself to be interpreted as an initial free energy measure. Bennett argued that information itself has a kind of 'fuel value' or free energy content. Following Feynman, we take this and run with it. In what follows, 'channel capacities' R_i will be used to build higher-order, iterated, free energy measures following from the statistical mechanical partition function.

That is, this formulation leads directly to a partition function definition of an *iterated* free energy F (Feynman 1998; Landau and Lifshitz 2007; Pettini 2007) as

$$\rho(F/\langle g\rangle) = \rho(R_1/g_1(Z_1)) + \rho(R_2/g_2(Z_2)) \tag{5.3}$$

where $\langle g\rangle$ is some appropriate 'average temperature' that we take here as

$$\langle g\rangle \equiv \sqrt{g_1(Z_1)g_2(Z_2)} \tag{5.4}$$

We next impose a first-order model, $g(Z) = g \times Z$, where the constant g is an *affordance* index that determines how well the resource rate Z can actually be utilized, and, for convenience, take $\rho(x)$ as the Boltzmann Distribution, solving for F as

$$F = -\left(\sqrt{g_1 Z_1 g_2 Z_2}\right) \ln\left(e^{-\frac{R_1}{g_1 Z_1}} + e^{-\frac{R_2}{g_2 Z_2}}\right) \tag{5.5}$$

Dynamics are then defined in second order at the nonequilibrium steady state according to an Onsager nonequilibrium thermodynamics model (de Groot and Mazur 1984)

$$\begin{aligned} S &\equiv -F + Z_1 dF/dZ_1 + Z_2 dF/dZ_2 \\ dZ_j/dt &\propto \partial S/\partial Z_j = 0, \; j = 1, 2 \end{aligned} \tag{5.6}$$

where S, as the Legendre Transform of F (Pettini 2007), is an entropy-analog, and we are able to omit the 'diffusion coefficient' in the second line of Eq. (5.6).

Solving the two relations $\partial S/\partial Z_j = 0$ for the R_j leads to three solution sets for the nonequilibrium steady state (nss)—i.e., homeostatic/allostatic—conditions. The physically nontrivial one is the *equipartition*

$$
\begin{aligned}
&\frac{R_1}{g_1 Z_1} = \frac{R_2}{g_2 Z_2} \equiv X \\
&Z_2 = \left(\frac{g_1 R_2}{g_2 R_1}\right) Z_1 \\
&Z \equiv (Z_1 + Z_2) = \left(\frac{1}{X}\right)\left(\frac{R_1}{g_1} + \frac{R_2}{g_2}\right)
\end{aligned}
\tag{5.7}
$$

1. Note that pathological, values of R_1, g_1 and Z_1 can force Z_2—the regulatory system supply demand—to grow so large that it becomes unsustainable. That is, homeostasis/allostasis of a cognitive/control dyad can be driven to failure by sufficient 'environmental' stress.
2. The 'total energy rate' Z can be *lowered* by raising the 'affordance' measures g_j.

This relatively simple example provides a useful starting point for significant generalizations, at the expense of some considerable and often nontrivial conceptual and formal development.

5.4 Extending the Model I

As Stanley et al. (2000) describe, empirical evidence has been mounting that supports the intriguing possibility that a number of systems arising in disciplines as diverse as physics, biology, ecology, and economics may have certain quantitative features that are intriguingly similar. These properties are conventionally grouped under the headings of 'scale invariance' and 'universality' in the study of phase transition.

Here, via the equipartition implied by Eq. (5.7), we *impose* a quasi-scale invariant model and explore resulting patterns of phase transition affecting and afflicting homeostasis and allostasis under 'stress'. The 'quasi' arises because one cannot simply abduct the dynamics of physical process to the dynamics of cognition, most particularly, biocognition. At the very least, information sources 'dual' to cognitive phenomena (Wallace 2022) are not micro-reversible, displaying, as mentioned above, directed homotopy.

An inverse argument generalizes the nss equipartition derived above across a full set of probability distributions likely to characterized biological, institutional, and machine cognition.

Fix the $R_j/(g_j(Z_j)) \equiv X$, and assume there are $n \geq 2$ contending/cooperating agents, taking n as an integer. (In analogy with physical theory, we might actually need to take $n \to V(n)$ for some monotonic increasing function of n.)

We are, again, envisioning at least an individual enmeshed in a social environment. Then, on this basis, calculate the most general possible form of the corresponding average R.

The relations of interest for possible underlying probability distributions are then taken according to the *first-order quasi-scale invariant* expressions

$$\rho\left(\frac{FX}{R(R_1, \ldots, R_n)}\right) = n\rho(X)$$
$$F = \left(\frac{\rho^{-1}(n\rho(X))}{X}\right) R(R_1, \ldots, R_n) \qquad (5.8)$$

where, again, n is an integer, and the appropriate functional form of $R = R(R_1, \ldots, R_n)$ must now be found.

See Khaluf et al. (2017) for discussion of 'ordinary' scale invariance where $\rho(\lambda x) = Q(\lambda)\rho(x)$ for some monotonic increasing function $Q(\lambda)$.

In general, it may be necessary to replace n by some appropriate monotonic increasing function $V(n)$, but for algebraic simplicity we will continue with the simple, first-order model.

Equation (5.6) is reexpressed in terms of the R_j, imposing the nonequilibrium steady state condition

$$\partial R_j/\partial t \propto \partial S/\partial R_j \equiv 0 \qquad (5.9)$$

for all possible R_j, so that the $R_j/(g(Z_j))$ are indeed fixed at the same value across all possible probability distributions.

What functional forms of R make this true?

The argument is not quite trivial but straightforward. *Requiring* $\partial S/\partial R_j = 0$ in Eq. (5.6), generalizes to a full class of 'averaging' functions $R(R_1, \ldots, R_n)$, given proper dimensional extension of S.

For n dimensions, a direct solution by induction is

$$R = R(R_1, R_2, \ldots, R_n) = f\left(\frac{R_2}{R_1}, \ldots, \frac{R_n}{R_1}\right) R_1 \qquad (5.10)$$

which applies to both arithmetic and geometric averages. This is a simplification from a more elaborate treatment that would add considerable mathematical overhead to an already burdened development.

We will reconsider this problem, and that of Eq. (5.4), from a different and perhaps deeper perspective below.

Here, we carry the argument further, treating Eq. (5.8) as an explicit function of X at nonequilibrium steady state, and extending the Onsager analysis:

$$\rho\left(\frac{F(X)X}{R(X)}\right) \equiv \rho(Q) = n\rho(X)$$
$$F(X) = \left(\frac{\rho^{-1}(n\rho(X))}{X}\right)R(X) = \frac{Q(n,\ X)}{X}R(X)$$
$$S \equiv -F(X) + X\frac{d}{dX}F(X)$$
$$dX/dt \propto \partial S/\partial X = 0$$
$$R(n,\ X) = \frac{X(C_1X + C_2)}{Q(n,\ X)} \tag{5.11}$$

where in the third line we again invoke the Legendre Transform to define an 'entropy' in terms of a 'free energy'.

Again, it may be necessary to replace the integer n in the first line of Eq. (5.11) by an appropriate monotonic increasing function $V(n)$. In physical systems, typically, $V(n) \sim n^{\alpha}$, $\alpha > 0$.

Higher order 'Onsager-like' (de Groot and Mazur 1984) expressions for the entropy-analog may well be needed, for example,

$$S \equiv -F + \sum_j \varepsilon_j X^j d^j F/dX^j \tag{5.12}$$

for appropriate ε_j. Such expressions may be system-specific, again leading to the second relation in Eq. (5.11), but producing more involved expressions for $F(X)$.

Below, we extend the argument to cases where $dX/dt \propto dS/dX = f(X(t)) \rightarrow 0$ as t increases, introducing 'frictional' delays.

Here, the essential point is that *calculating $Q(n, X)$ allows finding critical values $X_C(n)$ under different probability distributions and 'Onsager-type'—or other—models*. Again, scaling with n may not be the simple linear model used here.

Thus a direct, comprehensive, approach to system stability, via this approach, is to 'simply' demand that $Q(n,\ X) \neq 0$ be real-valued, given that $\rho(x)$ is a probability distribution on $0 \leq x \leq \infty$.

For, in order, Boltzmann, k-order Lomax, and Cauchy Distributions—instantiating the non-zero requirement—and sequentially, the Rayleigh, k-order Erlang and k-order Chi Square Distributions,

$$X_C = \ln(n)$$
$$X_C = n^{1/(k+1)} - 1$$
$$X_C = \sqrt{n-1}$$
$$X_C = \sqrt{-W\left(m,\ -\frac{\mathrm{e}^{-1}}{n^2}\right)}$$
$$X_C = -(k-1)\,W\left(m,\ -\mathrm{e}^{-1}\left(\frac{1}{n}\right)^{\frac{1}{k-1}}\right)$$

$$X_C = -(k-2)\, W\left(m, -\frac{(k-2)^{\frac{k}{k-2}} \left(e^{\frac{1}{k-2}}\right)^2}{\left((k-2)^{\frac{1}{k-2}}\right)^2 \left(n^{\frac{1}{k-2}}\right)^2 e^{\frac{k}{k-2}} (k-2)}\right) \tag{5.13}$$

where all scale parameters have been set equal to one and, in general, one might expect $n \to V(n)$ for some appropriate increasing function. Again, detailed development would add more mathematical overhead to an already overburdened argument.

Note that:

- The second condition is most easily proved by direct substitution.
- $W(m, x)$ is the Lambert W-function of order m that satisfies the relation

$$W(m, x) \exp[W(m, x)] = x$$

  W is real-valued only for $m = 0, -1$ and $x > -\exp[-1]$.
- Equation (5.13) is strictly equivalent to the condition that there is no solution to $\rho(Q) = n\rho(X)$ if $n\rho(X) > \max_{x \geq 0} \rho(x)$. If the distribution is nonmonotonic and there is a solution, there will be more than one solution form. The first three relations of Eq. (5.13) are associated with monotonic distributions having mode zero.

We are concerned with the *rising* branches representing cooperation between system subcomponents, displaying in Fig. 5.2, patterns of limiting behavior with increasing n for, respectively, the Rayleigh, second-order Erlang, and fourth-order Chi Square Distributions. These are nonmonotonic. The regimes above the top branch of the 'C' are stable for cooperating systems.

The essential point for this analysis is that, like $X_C = \ln(n)$ for the Boltzmann Distribution, the 'cost' in X for cooperating modes does not increase linearly with rising n. For the Boltzmann Distribution model, doubling n, i.e., pairing each cognitive submodule with a regulator, increases the rate of 'resource' demand as $\ln(n) \to \ln(2n) = \ln(n) + \ln(2)$.

For cooperating cognition/regulation dyads, the 'cost' in units of X does not increase in proportion to the shift $n \to 2n$. From the perspective of evolution under selection pressure, this seems a small price to pay for a grand increase in stability-of-function.

Again, here, the $m = -1$ branches can be taken as indexing cooperative coalitions, while the $m = 0$ branches represent contending system dynamics under rising n, which is another matter.

After some algebra, results similar to Fig. 5.2 are found for the Levy and Weibull distributions, although the Levy is difficult to parse.

We can now extend the 'energy' argument of Eq. (5.7), recalling that, for all j, $R_j/g_j(Z_j) \equiv X$, the equipartition value. Applying the arguments above, we set $X \equiv X_C(n)$ from Eq. (5.13) for the appropriate underlying probability distribution.

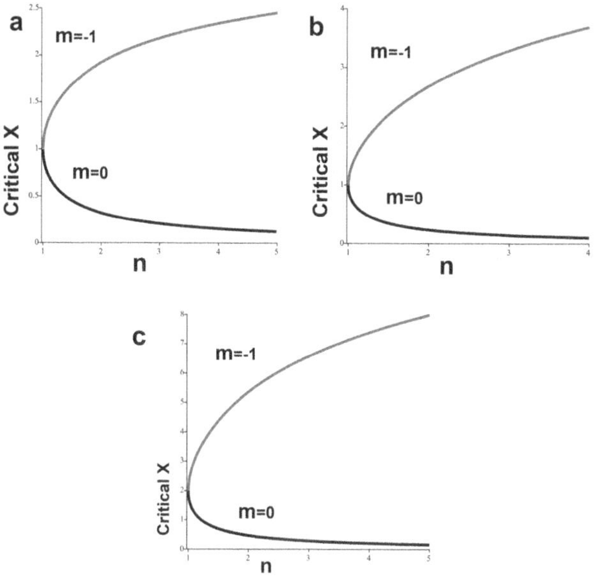

Fig. 5.2 Nonmonotonic distributions. For structures dominated by **a** Rayleigh, **b** second-order Erlang, and **c** fourth-order Chi Square Distributions, there are two solution forms and the regions within the 'C' are unstable. The systems are partitioned into $m = -1$ 'cooperating' and $m = 0$ 'contending' modes for n interacting components. The cooperating modes are stable only above the 'C'. Contending modes collapse with rising n. The minimum stability 'cost' in X for cooperating modes, defined by the tops of the 'C', however, does not increase linearly with rising n

Then

$$
\begin{aligned}
g_j(Z_j) &= \frac{R_j}{X_C(n)} \\
Z_j &= g_j^{-1}\left(\frac{R_j}{X_C(n)}\right) \\
Z_C &= \sum_j Z_j = \sum_j g_j^{-1}\left(\frac{R_j}{X_C(n)}\right)
\end{aligned}
\tag{5.14}
$$

For convenience, we assume a set of n identical systems, with fixed forms for the $g_j(Z_j) = g(Z_j)$ and fixed values of the $R_j = R$ under some particular probability distribution. Then

$$
Z_C = \sum_j Z_j = n g^{-1}\left(\frac{R}{X_C(n)}\right) \tag{5.15}
$$

Recall that, if g is monotonic increasing, so is g^{-1}.

For example, assuming a Boltzmann Distribution,

$$Z_C = ng^{-1}\left(\frac{R}{\ln(n)}\right) \tag{5.16}$$

which does not increase linearly with n by virtue of the division by $\ln(n)$ in g^{-1}.

Adopting the simple linear model $g(Z) \equiv g \times Z$,

$$\begin{aligned} \frac{R}{gZ_C} &= \frac{X_C(n)}{n} \\ Z_C &= \frac{n}{X_C(n)}\frac{R}{g} \end{aligned} \tag{5.17}$$

which makes intuitive sense.

Further, the first relation of Eq. (5.17) can be solved for n (or $V(n)$), generating necessary conditions for Z_C.

For example, under the Boltzmann Distribution,

$$n = \frac{-W\left(m, -\frac{R}{g\,Z_C}\right)}{\frac{R}{g\,Z_C}} \tag{5.18}$$

for $m = 0, -1$ in the Lambert W-function, imposing the necessary condition

$$\begin{aligned} \frac{R}{g\,Z_C} &\leq \exp[-1] \\ Z_C &\geq \frac{R}{g}\exp[1] \end{aligned} \tag{5.19}$$

From Eqs. (5.15)–(5.17), this is a general result for the linear model: That is, the same expression—with different right hand terms—is 'easily' found for the other probability distributions leading to Eq. (5.13).

5.5 Extending the Model II

The Basic Idea

The burdens of fog, friction, and adversarial intent characterize evolutionary selection pressures at every scale and level of organization of cognitive process, be it biological, social, institutional, machine, or their various composites.

Here, in address of these matters, we again assume the equipartition implied by Eq. (5.7) as a basic condition, and impose a kind of 'Born-Oppenheimer Approximation'

(Wikipedia 2025b) in which large-scale changes to the overall system are assumed to occur slowly enough that equipartition is 'almost always' sufficiently satisfied.

Under that basic assumption, friction is 'easily' introduced by extending the next-to-last relation of Eq. (5.11), so that

$$
\begin{aligned}
S &\equiv -F(X) + X\frac{d}{dX}F(X) \\
dX/dt &\propto dS/dX = Xd^2F/dX^2 = f(X) \\
F(X) &= \int\int \frac{f(X)}{X} dX dX + C_1 X + C_2
\end{aligned}
\tag{5.20}
$$

Here, we take

$$
\begin{aligned}
f(X) &= \beta - \alpha X(t) \\
X(t) &= \frac{\beta}{\alpha}(1 - \exp[-\alpha t]) \rightarrow \beta/\alpha
\end{aligned}
\tag{5.21}
$$

so that α becomes the (inverse) index of Clausewitzian friction, i.e., the delay in reaching the maximum possible value of X.

Then

$$
F(X) = \beta\,(X \ln(X) - X) - \frac{\alpha X^2}{2} + C_1 X + C_2 \tag{5.22}
$$

where C_1 and C_2 are important local boundary conditions that emerge 'naturally' from solving the differential equations. Solutions are indeed sensitive to these boundary conditions.

From Eq. (5.11), depending on the underlying probability distribution $\rho(X)$, recall that

$$
\begin{aligned}
R(n, X) &= \frac{X\,F(X)}{Q(n, X)} \\
\rho(Q(n, X)) &\equiv n\rho(X)
\end{aligned}
\tag{5.23}
$$

where Q is characteristic of that distribution. Again, in general, $n \rightarrow V(n)$.

Assuming monotononic Boltzmann and nonmonotonic Rayleigh and second order Erlang Distributions, i.e., $\rho(x) = \exp[-x],\ x\exp[-x^2/2],\ x\exp[-x]$, the $Q(n, X))$ are respectively,

$$
\begin{aligned}
& X - \ln(n) \\
& \frac{X\,\mathrm{e}^{-\frac{X^2}{2}}}{\sqrt{-\frac{X^2 \mathrm{e}^{-X^2}}{W\left(m, -X^2 n^2 \mathrm{e}^{-X^2}\right)}}} \\
& - W\left(m, -nX\,\mathrm{e}^{-X}\right)
\end{aligned}
\tag{5.24}
$$

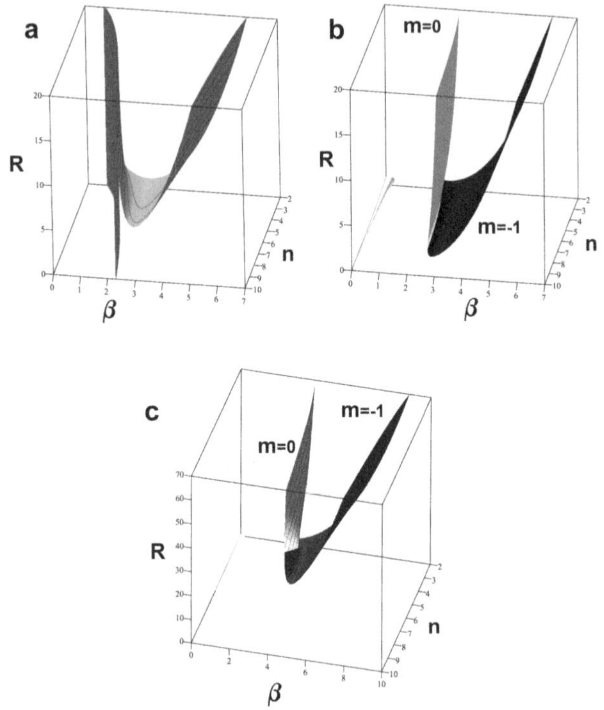

Fig. 5.3 $R(n, \beta)$, setting $\alpha = 1$ and, in the expression for $F(X)$, $C_1 = 1$, $C_2 = 2$. **a** Boltzmann, **b** Rayleigh, **c** second order Erlang distributions. Narrow troughs are bounded by unattainable demands for bandwidth. The nonmonotonic Rayleigh and Erlang distributions have two solutions, corresponding to the $m = 0, -1$ branches of the Lambert W-function

where $W(m, x)$ is the Lambert W-function of order $m = 0, -1$.

Recall that F is defined by $dS/dX = Xd^2F/dX^2 = dX/dt = f(X)$ so that F is given as the last relation of Eq. (5.20).

Fixing $\alpha = 1$ in the expression for F based on the exponential relation $f(X) = \beta - \alpha X$, with $C_1 = 1$, $C_2 = 2$, then $R(n, \beta)$ is as in Fig. 5.3a–c. Note the two solutions for the Rayleigh and Erlang Distributions, depending on the order of the Lambert W-function as indicated.

Recall that R is an average channel capacity for control across the full system: This is not a trivial matter. Here, relatively narrow troughs are bracketed by high—essentially unattainable—demands for bandwidth under friction measured by α.

Fog and friction

In addition to the burden of Clausewitzian 'friction', inherent to the relation $dX/dt = f(X(t)) \rightarrow 0$, the effect of 'fog-of-war' can be explored by expanding the expression for dX/dt in Eq. (5.20) as a stochastic differential equation

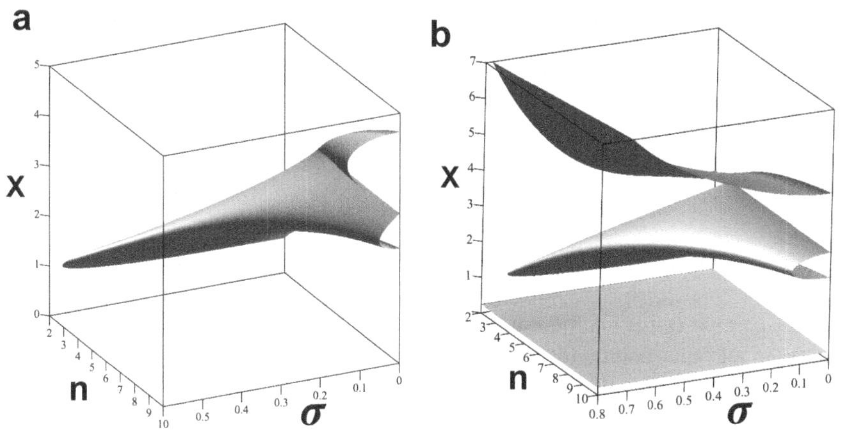

Fig. 5.4 Effect of increasing stochastic 'fog' σ on equipartition value X for the Boltzmann distribution assuming **a** Arrhenius and **b** exponential models for $dX/dt = f(X)$ under the nonequilibrium steady state condition $< dR_t >= 0$. Here, $\beta = 3$, $\alpha = 1$, $C_1 = -1$, $C_2 = 3$ in F. As n and σ increase, the solution sets are destabilized

$$dX_t = f(X_t)dt + \sigma X_t dB_t \tag{5.25}$$

where dB_t is taken as Brownian white noise, and the last term represents 'ordinary' volatility, here in the context of two models, the 'exponential' and the 'Arrhenius', given a Boltzmann Distribution and n cognitive modules with the equipartition assumption on X.

The exponential model follows Eqs. (5.21) and (5.22). For the Arrhenius model,

$$X(t) = \beta \exp[-\alpha/t]$$

$$dX/dt = f(X) = \frac{X}{\alpha}\,(\ln(X/\beta))^2$$

$$F(X) = \frac{\ln\left(\frac{X}{\beta}\right)^2 X^2}{2\alpha} - \frac{3X^2 \ln\left(\frac{X}{\beta}\right)}{2\alpha} + \frac{7X^2}{4\alpha} + C_1 X + C_2 \tag{5.26}$$

Assuming the Boltzmann Distribution—so that $Q = X - \ln(n)$ in Eq. (5.23), where n is the number of linked modules—we calculate the nonequilibrium steady state conditions $< dR_t >= 0$ for both examples using the Ito Chain Rule (Cyganowski et al. 2002). The resulting expressions are relatively simple, but too lengthy for the author's LaTex compiler.

For both forms of F, exponential and Arrhenius, $\alpha = 1$, $\beta = 3$, $C_1 = -1$, $C_2 = 3$. Figure 5.4 shows the resulting nss solution sets, expressed as $X(n, \sigma)$. As n and σ increase, the solution sets are destabilized.

It is interesting to repeat this calculation for the second-order Erlang Distribution, representing a process of two sequential, identical, Exponential Distributions, e.g., two-fold staged decision, using the convolution integral method described in the Mathematical Appendix. Then

$$\rho(x) = x \exp[-x]$$
$$Q(n, X) = -W\left(m, -\frac{nX}{e^X}\right) \tag{5.27}$$

where, again, $W(m, x)$ is the Lambert W-function of order m.

Taking all other parameter values exactly as in Fig. 5.4, Fig. 5.5 again represents the nonequilibrium steady state solution sets $X(n, \sigma)$ for the Arrhenius and exponential models defining $dX/dt = f(X)$. The solution sets become counter intuitively convoluted with increasing 'noise' σ.

These examples show how Clausewitzian 'friction' defined by α in the expressions for $dX/dt = f(X)$ becomes idiosyncratically synergistic with 'fog' uncertainties characterized by σ. That is, what appear to be 'slight' differences between frictional dynamics—different forms of delay in response to perturbation—can have amplified impact on the dynamics of uncertainty in response to perturbation. Similar results will afflict all probability models across any modular conformations and forms of the friction relation $f(X)$.

These are not encouraging results, as any embodied cognitive system, dynamically tuned modular or otherwise, will always encounter rising rates of 'noise' in chaotic, unsecured or adversarial environments under necessarily delayed response to such perturbations.

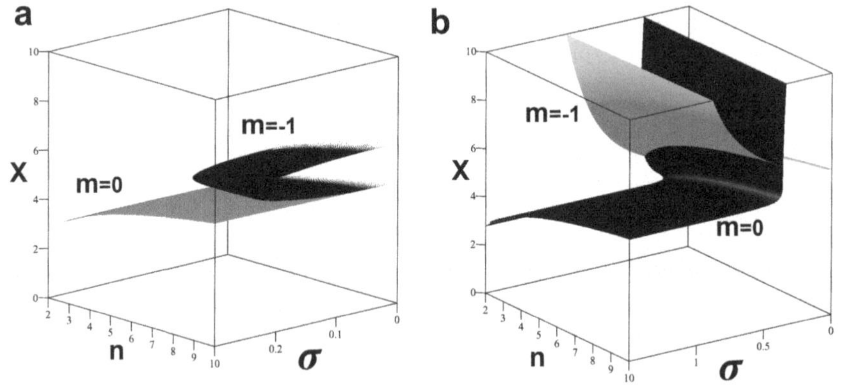

Fig. 5.5 Analog to Fig. 5.4 for the second-order Erlang distribution representing the serial composition of two Boltzmann distributions. **a** Arrhenius. **b** Exponential. What appear to be 'slight' differences between frictional dynamics appear to have amplified effect in synergism with uncertainty

The statistical tools derived from this general approach will likely be fully as flexible for the study of cognitive phenomena as have been the 'ordinary' regression equations used in the analysis of more usual observational and experimental data, in particular providing benchmarks against which to compare observed patterns.

Cognition, cognitive dynamics—and their inherent failure modes—are hard to understand and remediate, and their effective address requires new methods and approaches.

What seems to emerge from the synergism between uncertainty and frictional delay in response to that uncertainty in cognitive dynamics is the dire need for analogs to the long-evolved innate and adaptive immune systems, i.e., a basic, rapid 'implanted' response, followed by a more detailed 'learned' response.

5.6 Groupoid Symmetry-Breaking Phase Transitions

There is yet more to be teased out from behind the veils of Clausewitzian Fog-and-Friction. So far, we have treated 'underlying probability distributions' as the central, fixed star in a kind of planetary dynamic, leading to the methodological harvests of Eq. (5.13) and what follows from the network relation $\rho(Q) = n\rho(X)$ (recalling that, in general, $n \to V(n) \sim n^{\alpha}$).

What happens when $\rho(X)$ is, itself, perturbed by fog and friction?

Most simply, this can be found by calculating the nonequilibrium steady state

$$\langle d\rho_t \rangle = 0$$

based on the generalized 'frictional' stochastic differential equation

$$dX_t = f(X_t)dt + \sigma h(X_t)dB_t \tag{5.28}$$

using the Ito Chain Rule.

The relation defining the nss is 'easily' found as

$$\begin{aligned}\sigma^2 &= -\frac{2f(X)\left(\frac{d}{dX}\rho(X)\right)}{h(X)^2\left(\frac{d^2}{dX^2}\rho(X)\right)}\\ &= -\frac{2X\frac{d^2}{dX^2}F(X)\left(\frac{d}{dX}\rho(X)\right)}{h(X)^2\left(\frac{d^2}{dX^2}\rho(X)\right)}\end{aligned} \tag{5.29}$$

where the second form follows from the second expression in Eq. (5.20), i.e., $dX/dt \propto dS/dX = Xd^2F/dX^2 = f(X)$.

Equation (5.29) generates solution set equivalence classes $\{\sigma, \ X\}$ that can be seen as arising from the functions $X(\sigma)$ or $\sigma(X)$.

1. For reasons that will become more salient, if not intrusive, we are principally concerned with these equivalence classes rather than with the functions themselves.
2. The term

$$-\frac{d\rho/dX}{d^2\rho/dX^2}$$

   may be interpreted as the scale of change or 'characteristic length' relating how sharply the density changes versus its curvature at a point. It is analogous to, but different from, the classic 'hazard rate' at which signals are detected. For the Exponential Distribution $m \exp[-mX]$, the hazard rate is m and the characteristic length $1/m$.
3. For convenience, we adopt 'simple volatility', i.e., $h(X) = X$, with $f(X)$ as the exponential and Arrhenius models from above for, respectively, the Boltzmann and Second Order Erlang Distributions, and plot $\sigma(X)$ to generate Fig. 5.6.

Again, the second-order Erlang represents the convolution of two identical exponential distributions, akin to replacing 'mission command' initiative with a two-stage 'detailed command' protocol in military or business enterprise (Wallace 2025).

The distributions are, respectively, monotonic and nonmonotonic. The results, in Fig. 5.6, are **a** Boltzmann-exponential, **b** Boltzmann-Arrhenius, **c** Erlang-exponential, and **d** Erlang-Arrhenius. Again,

$$\begin{aligned}\rho(X) &= \exp[-X], \;\; X\exp[-X] \\ f(X) &= \beta - \alpha X, \;\; (X/\alpha)(\ln(X/\beta))^2\end{aligned} \tag{5.30}$$

Here, $\alpha = 1, \;\; \beta = 3$ in both expressions for $f(X)$.

These distinct solution sets, representing markedly different fundamental interactions between internal probabilities, 'friction' from α and β, and 'fog' from σ, can be grouped together as equivalence classes even as the internal parameters defining $\rho(X)$ and $f(X)$ are varied.

As the Mathematical Appendix of the previous chapter explores in some detail, equivalence classes lead to an important generalization of elementary ideas of group symmetry, i.e., groupoids, for which products are not defined between all possible element pairings (Weinstein 1996).

The simplest groupoid is the disjoint union of a set of groups. Imagine two bags containing the same number of identical ball bearings, each of which can be characterized by the three-dimensional rotation group. Take a ball bearing from one bag, make a good indentation in it with a hard-metal punch, and put it back in the original bag. Now the disjoint union of three-dimensional rotation groups in one of the bags has been broken. One bearing is now symmetric only via rotation about an axis defined by the indentation.

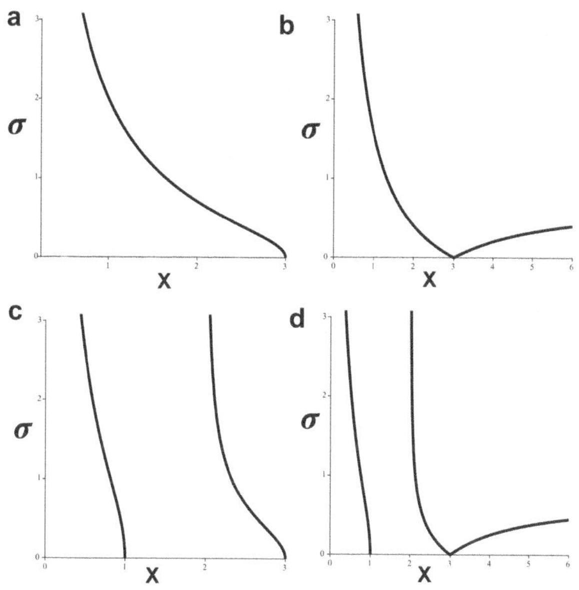

Fig. 5.6 Assuming 'simple volatility' $h(X) = X$, with $f(X)$ as the exponential and Arrhenius models, generates equivalence class solution sets **a** Boltzmann-exponential, **b** Boltzmann-Arrhenius, **c** Erlang-exponential, and **d** Erlang-Arrhenius. For all, $\alpha = 1,\ \beta = 3$

The physical act of hitting one ball bearing with a hard-metal punch has triggered a groupoid symmetry-breaking phase transition, analogous to—but different from—group symmetry-breaking phase transitions like melting a snow crystal, evaporating a liquid, or burning a diamond.

Each panel of Fig. 5.6 defines a particular set of equivalence classes that, via the somewhat mind-bending arguments of the Appendix to the last chapter, defines a groupoid. Changes within fog/friction/probability patterns under the perturbations of external circumstance—like the hard-metal punch on the ball bearing—trigger punctuated groupoid symmetry-breaking phase transitions in the kind of systems we have studied. That is, sufficiently challenged cognitive structures will experience sudden shifts among the panels of a vastly enlarged version of Fig. 5.6.

A more extensive discussion of groupoid symmetry-breaking phase transitions by the Perplexity AI system is found in the Appendix.

Cognition, even with regulation, is a castle made of sand.

5.7 Iterated System Temperature

Subsystems of living organisms consume metabolic free energy (MFE) at markedly different rates. For example, among humans, Heart/Kidneys, Brain, and Liver typically require, respectively 440, 240, and 200 KCal/kg/day of MFE, while resting Skeletal Muscles and Adipose Tissue require only 13 and 4.5 Kcal/kg/day (Wang et al. 2010, Table 4).

These marked differences, however, are encapsulated within a 'normal body temperature' of $\approx 98.6\,^{\circ}$F.

This is not as simple as it seems, or even as treated in the first iteration above of Eqs. (5.4) and (5.10).

Here, we have, via an 'equipartition' argument, condensed closely interacting nonequilibrium steady state subcomponents of a larger enterprise in terms of the quantity X from Eqs. (5.20)–(5.29). Taking the perspective of the previous section, more can be done, leading in particular to the definition of an 'iterated temperature' construct seemingly akin to the normal body temperature.

The essential assumption is, again, that, under the stochastic burden of Eq. (5.28), the basic underlying probability distribution is preserved in a nonequilibrium steady state $\langle d\rho_t \rangle = 0$, leading to the second form of Eq. (5.29). Then the 'free energy' construct $F(X)$ is expressed as

$$F(X) = \int \int -\frac{\sigma^2 h(X)^2 \left(\frac{d^2}{dX^2}\rho(X)\right)}{2X\left(\frac{d}{dX}\rho(X)\right)} dX dX + C_1 X + C_2 \tag{5.31}$$

Recall now, from above, the embodied Rate Distortion Control Theory partition function argument defining 'free energy' F in terms of a temperature analog g as

$$\rho(F/g) = \int_0^\infty \rho(R/g) dR = g \tag{5.32}$$

where, here, g, has become an iterated function of X, the equipartition energy measure. From here on, we will write this as G^* to differentiate it from the 'simple' temperature analog of previous studies.

That is, here, we are interested in the solution to the relation

$$\rho(F/G^*) = G^*(X) \tag{5.33}$$

given the nss condition $\langle d\rho_t \rangle = 0$, i.e., the quasi-preservation of the underlying basic probability distribution of the overall system.

We are, in a sense, reprising the calculations of Eqs. (5.4) and (5.10) from a different perspective.

More specifically, we contrast a simple one-stage 'mission command' system based on a Boltzmann Distribution with a two-stage 'detailed command' process that passes decisions across two sequential Boltzmann stages, giving a process described overall by a second-order Erlang Distribution. Under the impact of 'noise' defined by Eq. (5.28), each 'cultural' form attempts homeostatis, the preservation of its inherent nature, i.e., $\langle d\rho_t \rangle = 0$.

Thus we, respectively, set $\rho(x) = \exp[-x]$, $x \exp[-x]$ and assume ordinary volatility $h(X) = X$. Then, again via the Ito Chain Rule calculation (Cyganowski et al. 2002),

$$
\begin{aligned}
F_B(\sigma, X) &= \frac{1}{12}\sigma^2 X^3 + C_1 X + C_2, \\
G_B^* &= \frac{-F_B}{W(m, -F_B)} \\
F_E(\sigma, X) &= \frac{\sigma^2}{4}\left(\frac{X^3}{3} - X^2 - 2\ln(X-1)(X-1) - 2 + 2X\right) \\
&+ C_1 X + C_2, \\
G_E^* &= \frac{-F_E}{2\,W\left(m, \pm\frac{\sqrt{F_E}}{2}\right)}
\end{aligned}
\tag{5.34}
$$

where $W(m, x)$ is the Lambert W-function of orders $m = 0, -1$. The $G^*(X, \sigma)$ expressions are then the iterated temperatures of the entire system under the reduced equipartition defining X across subsystems.

Here, we adopt the negative-$\sqrt{F_E}$ form of the W-function in G_E^*, as the positive form generates only negative temperatures that, in physical systems, are associated with grossly unstable 'excited states'.

Figures 5.7a and b show the two forms of $G^*(X, \sigma)$, setting $C_1 = -1$, $C_2 = 3$, with $m = 0, -1$ as indicated.

Sufficient 'noise' σ drives both systems into zones of instability where the relation $\langle d\rho_t \rangle = 0$ fails. This occurs, via the properties of the W-function, for, respectively, $F_B(X, \sigma) > \exp[-1]$ and $\sqrt{F_E}/2 > \exp[-1]$ or $F_E < 0$. The contrast between stable zones is noteworthy: The Detailed Command model is at best thinly stable, in comparison with our version of Mission Command.

Again, each mode of underlying 'culture', when stressed, attempts homeostasis as the nonequilibrium steady state $\langle d\rho_t \rangle = 0$.

It is 'not difficult' to extend these results across orders of iteration of the underlying Boltzmann Distribution via a k-stage 'cancer initiation' model based on the Gamma Distribution

$$
\rho(m, k, X) = \frac{m^k X^{k-1} e^{-mX}}{\Gamma(k)}
\tag{5.35}
$$

where $\Gamma(x)$ is the Gamma Function that generalizes the factorial function for integers.

Then the nss condition $\langle d\rho_t \rangle = 0$, under the Ito Chain Rule, gives

$$
\begin{aligned}
\frac{d^2}{dX^2}F(X) &= \frac{\sigma^2\left(m^2X^2 - 2mXk + 2mX + k^2 - 3k + 2\right)}{2mX - 2k + 2} \\
G^* &= -\frac{mF}{kW\left(n, -\frac{\Gamma(k)^{\frac{1}{k}} F^{\frac{1}{k}}}{k}\right)}
\end{aligned}
\tag{5.36}
$$

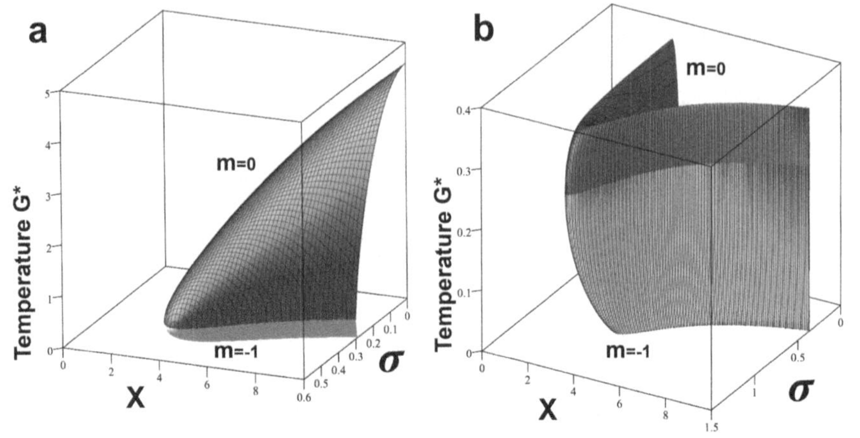

Fig. 5.7 Iterated temperature $G^*(X, \sigma)$ for the **a** Boltzmann and **b** second-order Erlang Distributions with $C_1 = -1$, $C_2 = 3$ in F. Both systems become unstable, respectively, for $F_B > \exp[-1]$ and $\sqrt{F_E}/2 > \exp[-1]$ or $F_E < 0$. The stable zone of the Detailed Command example is consequently highly compressed and stunted

The first relation can easily be solved as $F = \sigma^2 H(m, k, X) + C_1 X + C_2$ in an expression too long to be written on this page.

We can conceptually simplify matters slightly by generating a three-dimensional 'criticality surface' $\{k, \sigma, X\}$ *across the decision level index* k by (1), using the necessary condition for real-valued Lambert W-functions of orders $n = 0, -1$, and (2), setting $m = 1$, while varying the boundary conditions C_1, C_2. This is done via the second expression of Eq. (5.36) as

$$\frac{\Gamma(k)^{\frac{1}{k}} F^{\frac{1}{k}}}{k} = e^{-1}$$
$$F = \frac{k^k e^{-k}}{\Gamma(k)} \tag{5.37}$$

The result is shown in Fig. 5.8a and b for two boundary condition pairs: $\{C_1 = -1, C_2 = 3\}$ and $\{C_1 = 1, C_2 = 2\}$.

One possible inference from this development is that the approach we have taken permits development of a very rich body of statistical tools for data analysis. Like all such, however, matters of error estimates and basic validity provide a correspondingly rich range of possible devils-in-the-details.

In the Chapter Mathematical Appendix we explore a 'higher iteration' in the detailed command structure, i.e., doubling the two-level dynamic based on the second order Erlang Distribution. The picture is of an initial two-stage 'tactical' decision level, followed by a second two-stage 'operational' or 'strategic' command level. The results are not reassuring.

It is important to note that other selection pressures may operate. That is, Eq. (5.31) assumes the most essential matter is the preservation of the underlying probability distribution through the nonequilibrium steady state $\langle d\rho_t \rangle = 0$. What if environmental selection pressure enforces the preservation of some other quantity, for example a 'Weber-Fechner' sensory signal detection mode $Q(X) = \ln(1 + X)$ instead of $\rho(X)$? Then $\langle dQ_t \rangle = 0$ under the Ito Chain Rule leads to

$$
\begin{aligned}
F_Q(\sigma, X) &= \frac{\sigma^2}{2}\left(\frac{X^2}{2} - \ln(1+X)(1+X) + 1 + X\right) + C_1 X + C_2 \\
G_B^* &= \frac{-F_Q}{W(m, -F_Q)} \\
G_E^* &= \frac{-F_Q}{2\,W\left(m, -\frac{\sqrt{F_Q}}{2}\right)}
\end{aligned}
\tag{5.38}
$$

Taking $C_1 = -1,\ C_2 = 3$ again produces something much like Fig. 5.7.

We can carry things a further step in terms of a 'collective cognition rate' defined by the basic underlying probability distribution of the overall multi-modular system, given the Rate Distortion Control Theory model of Fig. 5.1. The essential point is to define 'system temperature' as G^* from Eq. (5.34) or (5.38). Then, given the distribution form ρ, and a basic Data Rate Theorem stability limit R_0, the cognition rate, in analogy with a chemical physics reaction rate approach (e.g., Laidler 1987) becomes

$$
L \equiv \frac{\int_{R_0}^{\infty} \rho(R/G^*)dR}{\int_0^{\infty} \rho(R/G^*)dR} \tag{5.39}
$$

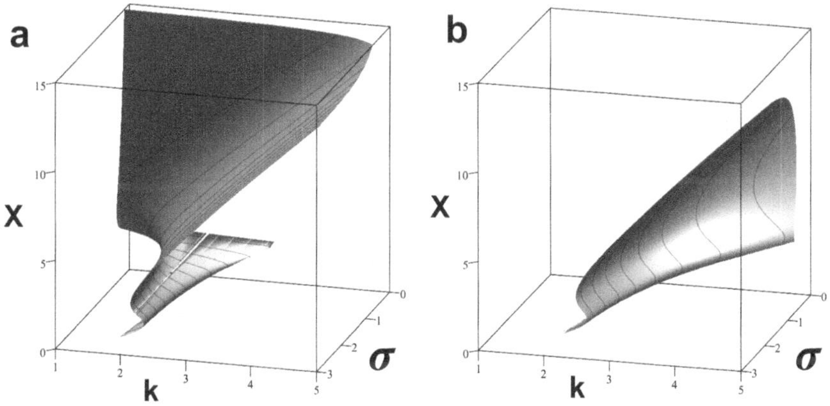

Fig. 5.8 Criticality surfaces from Eq. (5.37) for two boundary condition sets: **a** $\{C_1 = -1,\ C_2 = 3\}$ **b** $\{C_1 = 1,\ C_2 = 2\}$

For the particular Boltzmann and second-order Erlang Distributions studied here, Fig. 5.7 and friends are transformed as

$$
\begin{aligned}
G^* &\rightarrow \exp[-R_0/G^*] \\
G^* &\rightarrow \exp[-R_0/G^*]\left(1 + \frac{R_0}{G^*}\right)
\end{aligned}
\tag{5.40}
$$

Details are left as an exercise. The general forms of $L(X, \sigma)$ are, however, similar to Fig. 5.7.

Note that the 'Boltzmann' part of Eq. (5.40) produces the classic inverted-U 'Yerkes-Dodson Effect' pattern of cognition rate versus 'arousal' X across a variety of plausible $\langle dQ_t \rangle = 0$ nonequilibrium steady state restrictions (Wallace and Fricchione 2024; Wallace 2025).

5.8 Discussion

There seem to be 'universality classes' for cognitive systems, determined by the nature of the underlying probability distributions, e.g., monotonic or nonmonotonic, and by the nature of their embedding environments. Other examples may become manifest across different cognitive phenomena, scales, and/or levels of organization.

In particular, the transcendental Lambert W-function is used to solve equations in which the unknown quantity occurs both in the base and in the exponent, or both inside and outside of a logarithm (Wikipedia 2025). Similarly, problems from thermodynamic equilibrium theory involving fixed heat capacities parallel the results of Fig. 5.2.

Indeed, Wang and Moniz (2017) make explicit use of the Lambert W-function in the study of classic physical system phase transitions, using arguments recognizably similar to those developed here for cognitive phase transition. Likewise, Roberts and Vallur (2014) use the W-function in their study of the zeros of the QCD partition function, another phase transition analysis.

The Perplexity AI system (Perplexity AI 2025) contrasts physical and cognitive phase transition models in these terms:

Key Differences

Feature	Physical phase transitions	Cognition/consciousness models (Wallace)
Domain	Physics (thermodynamics, statistical mechanics, QCD, etc.)	Information theory, control theory, neuroscience
Control parameter	Physical (temperature, chemical potential, etc.)	Information source uncertainty, resource rate
Phenomenology	Material states, order-disorder, symmetry breaking	Cognitive state shifts, institutional or group cognition, regulatory transitions
Mathematical focus	Partition function zeros, explicit criticality expressions	Rate of cognition, stability/instability, bifurcations in control flow
Role of branch structure	Multiple physical solutions/metastability	Transitions between cognitive regimes, oscillations, complex dynamics
Institutional/social model	Absent	Central: addresses institutions, collective cognition, network effects

Both physical and 'cognitive' theories are based on probability distribution arguments 'in which the unknown quantity occurs both in the base and in the exponent, or both inside and outside of a logarithm'. Cognitive phenomena, of course, act over much richer and more dynamic probability landscapes. Cognitive system probability models will, consequently, require sophisticated and difficult conversion to statistical tools useful in the analysis of observational and experimental data.

Khinchin (1949) explored in great detail the rigorous mathematical foundations of statistical mechanics, based on such asymptotic limit theorems of probability theory as the Central Limit and Ergodic Theorems. Here, we have outlined the need for the parallel—but strikingly different—development of a rigorous 'statistical mechanics of cognition' based on the asymptotic limit theorems of information and control theories. Such developments must, however, inherently reflect the directed homotopy and groupoid symmetry-breaking mechanisms inherent to cognitive phenomena.

There are important matters buried in this formalism:

- We study systems at various scales and levels of organization that are, by Fig. 5.1, fully *embodied* in the sense of Dreyfus (1972), so that their cognitive dynamics may be, or become, 'intelligent' in his sense.
- X is basically a 'rate' variate, and the existence of a minimum stable X_C for cooperating cognitive/regulation dyads is consistent with John Boyd's OODA Loop analysis (Osinga 2007) in which a central aim is to force the rate of confrontation

to exceed the operating limit of one's opponent, breaking homeostatic or allostatic 'stalemate'. Predator/prey cycles of poste and riposte come to mind.

- The cost of the pairing of cognitive with regulatory submodules that is necessary for even basic stability under dynamic environments scales less than linearly with the inherent doubling of modular numbers across a considerable range of basic underlying probability distributions likely to characterize embodied cognition. This appears to be a very general matter having important evolutionary implications across a variety of biological, social, institutional, and machine venues.
- Nonetheless, in spite of 'best case' assumptions and deductions, failure of nominally 'effective' cognition on dynamic environments is not a 'bug', but is an inevitable feature of cognitive process of any and all configurations, at any and all scales and levels of organization, including systems supposedly stabilized by close coupling between cognitive and regulatory submodules. See the Appendix for a surprisingly broad literature review by the Perplexity AI system.

The last point has particularly significant implications. A recent study of nuclear warfare command, control, and communications (NC3) by Saltini et al. (2025) makes the assertion

> Because...[strategic warning, decision support and adaptive targeting]... are independent and tightly coupled, the potential role of AI is not limited to a single application; instead it can simultaneously enhance multiple aspects of the NC3 architecture.

The introduction of AI entities—even when closely paired with draconian regulatory modules—cannot be assumed to significantly stabilize or enhance NC3, or any other critical systems. Placing NC3 architectures under AI control simply serves to shift, obscure, and obfuscate failure modes.

This being said, as a reviewer has noted, the approach advocated here has a number of limitations:

- The g and ρ families used here are simplifications. Real systems may deviate. Sensitivity analysis—σ and SDE's—helps, but does not establish universality.
- Validation has been minimal by design. Thresholds/phase transitions should be viewed as testable predictions, not established empirical facts.
- Results may not hold in adversarial, highly non-stationary, or strongly path-dependent settings without additional assumptions or new proofs.

Again, this being said, on the basis of our—and much similar—work, the Precautionary Principle should constrain the use of AI in such safety-critical domains as NC3.

Further work will be needed to fully characterize the effects of Clausewitzian fog and friction across the full set of probability distributions likely to underlie a wide spectrum of observed cognitive phenomena, but what emerges from this analysis is the ubiquity of failure-under-stress even for 'intelligent' embodied cognition, even when cognitive and regulatory modules are closely paired.

There is still No Free Lunch, much in the classic sense of Wolpert and MacReady (1995, 1997).

5.9 Appendix

Groupoid Symmetry-Breaking Phase Transitions

Generated by the Perplexity AI system.

[The Wallace work]... connects phase-transition-like phenomena in cognitive systems to groupoid symmetry-breaking, a mathematical generalization of group theory that better captures local, context-dependent symmetries. The idea is that, just as physical systems undergo sharp shifts in macroscopic behavior (phase transitions) due to symmetry breaking, so too might cognitive or social systems when information flows or constraints change.

Physical Theories and Groupoid Symmetry-Breaking

While standard phase transition theory (as in Ising models or Landau's symmetry-breaking) mainly employs group symmetries, deeper mathematical frameworks—specifically groupoids—have been applied in various modern extensions:

- *Topological Phases of Matter*: In condensed matter and statistical mechanics, generalized symmetries, including groupoids, have been used to classify distinct phases, especially where local order parameters or defects are prominent. This context is particularly relevant in systems with spatially varying or local symmetries whose transformations do not fit neatly into group actions.
- *Quantum Field Theory (QFT)*: Groupoids appear in describing moduli spaces of field configurations, local gauge freedoms, and anomalies. In systems with nontrivial topology or defects (domain walls, vortices), groupoid symmetries formalize the algebra of admissible local transformations, essential in phase transitions involving topological order.
- *String Theory and Higher Category Theory*: While classic string theory phase transitions (such as confinement/deconfinement or duality transitions) have historically relied on group-based symmetries, modern approaches—especially those employing higher category theory—increasingly use groupoids (and even higher groupoids) to encode brane intersections, generalized gauge symmetries, and dualities. The language of moduli stacks (groupoids up to homotopy) is prominent in string compactification, where phase structure is analyzed via changes in the topology or geometry of moduli spaces under symmetry breaking.

String Theory: Phase Transitions and Groupoids

- *Moduli Stacks and Groupoids:* String theory's moduli spaces of vacua and their degenerations (leading to phase transitions, e.g., conifold transitions) are often described by stacks, mathematically formalized as groupoids-in-spaces. Symmetry-breaking is then realized as a shift in the groupoid of symmetries acting on these moduli, with phase transitions corresponding to changes in the equivalence classes of these groupoid actions.

- *Generalized Geometric Langlands Program:* This mathematical framework, motivated by string dualities and topological field theories, also formalizes symmetry-breaking via groupoid and higher categorical structures, analyzing how categorical symmetries fragment or recombine at criticality.

Summary Table: Groupoid symmetry-breaking in physical phase transitions

Theory/area	Role of groupoids	Phase transition context
Condensed matter physics	Local symmetry structures	Topological and nonlocal order transitions
Quantum field theory	Moduli, defects, anomalies	Dualities, topological changes
String theory, moduli spaces	Moduli stacks, brane categories	Conifold transitions, geometry changes
Geometric langlands/topological QFT	Higher groupoids/categories	Category-theoretic symmetry breaking

This convergence of ideas is at the frontier of mathematical physics and complex systems theory—and is a notable area where abstraction and application in cognitive science, as per Wallace, connect deeply with string theory and advanced modern physics.

Literature

Symmetry Breaking and Gauge Theory-Groupoids and Groupoidification by Baez and others explore mathematical physics models in categorical/groupoid frameworks.

Moduli Stacks and Groupoids in String Theory in algebraic geometry and physics journals detail these structures for moduli space phase transitions.

Wallace's work can be found in Consciousness, Cognition and Crosstalk: The evolutionary exaptation of nonergodic groupoid symmetry breaking, Springer, New York, 2022.

Doubling-down on Bureaucracy

Moving from a single-level decision structure under some probability distribution $\rho_1(x)$ to a two-stage model based on that distribution is done via the convolution integral

$$\rho_2(x) \equiv \int_0^x \rho_1(u)\rho_1(x-u)du \tag{5.41}$$

characterizing a two-stage dispersal. One could, at the expense of mathematical difficulty, compound the first and second stages using different basic probability distributions. We restrict the argument to Boltzmann and second-order Erlang Distributions.

It is easy to show that $\exp[-x] \rightarrow x \exp[-x]$ under the transformation of Eq. (5.41).

Suppose the PLA's (or the current USA's) Cultural Revolution-era Stavka requires extra 'political' supervision via the compounding of two two-stage Erlang Distributions. Then Eq. (5.41) gives $x \exp[-x] \rightarrow \rho_4(x) = (x^3/6)\exp[-x]$. Taking ordinary volatility under Eq. (5.31),

$$
\begin{aligned}
F &= \frac{\sigma^2}{4}\left(\frac{X^3}{3} - 3X^2 - 6\ln(X-3)(X-3) + 6X - 18\right) + C_1 X + C_2 \\
\rho_4(F/G^*) &= G^* \\
G^* &= -\frac{F}{4\,W\left(m, -\frac{6^{1/4}F^{1/4}}{4}\right)} \\
L &= \frac{e^{-\frac{R_0}{G^*}}\left(6{G^*}^3 + 6R_0{G^*}^2 + 3R_0^2 G^* + R_0^3\right)}{6{G^*}^3}
\end{aligned}
\tag{5.42}
$$

The critical relation, from the Lambert W-function, is

$$
\frac{6^{1/4}F^{1/4}}{4} = e^{-1} \tag{5.43}
$$

G^* and the cognition rate L, in this model, have become exceedingly sensitive to the boundary conditions C_1 and C_2, far more than perhaps expected. In particular, Fig. 5.7b is pancaked even further. Indeed, for the boundary conditions of Fig. 5.7, only the $\sigma = 0$ result is real-valued: any 'noise' at all is fatal to system cognition.

In some contrast, setting $C_1 = 1$, $C_2 = 2$ leads to a bizarre stochastic resonance in which a minimum noise level is required for any signal detection. Grim details are left as an exercise.

To yet again beat the Devil, as it were, multiple levels of control seem decidedly counterproductive when confronted by fog, friction, and skilled adversarial intent.

The Consensus on 'Psychopathology' and Regulation

The Perplexity AI (2025) system comments on these matters as follows.

... [A] number of researchers in cognitive science, neuroscience, and information theory have independently converged on conclusions... that instability and even failure are inherent to embodied cognitive systems, and that pairing with regulatory modules (e.g., meta-cognitive or executive functions) is essential for any measure of reliable function. This view asserts that dysfunctions—ranging from hallucination to broad psychopathology—arise not simply as defects to be eradicated, but as inevitable features of complex, embodied cognition interacting with a highly variable environment [1–4].

Cognitive Instability and Regulation

Several studies in neuroscience have framed cognitive control as a process of reducing uncertainty and stabilizing decision-making under constant environmental variability. The architecture of human cognition relies on layered control networks (especially in the frontoparietal and anterior cingulate cortices), designed to regulate or compensate for the unreliable and noisy nature of sensory and internal cognitive processes. Failures in these regulatory modules can manifest as behavioral or perceptual instability, which is increasingly viewed... as a structural property of complex cognitive architectures [4, 1].

Embodied Cognition and Predictive Coding

Theories rooted in embodied cognition and predictive coding (notably in accounts of schizophrenia and hallucination) describe how the interplay of top-down (cognitive/regulatory) and bottom-up (sensory/environmental) processes is inherently precarious. In disorders like schizophrenia, symptoms arise from an imbalance between these streams: a failure in predictive regulation that is not merely pathological 'noise,' but an outcome of how predictive, embodied systems must operate in uncertain worlds. Such failures are not bugs to be entirely eliminated, but intrinsic risks of the embodied, probabilistic strategies that make cognition flexible and adaptive [2, 4].

Feature or Bug? Theoretical Perspectives

Contemporary researchers have proposed that 'noise' in cognition is a central, adaptive feature rather than an aberration. For example, in a recent review, the variability of neural sampling and thought is described as critical for exploration, adaptation, and learning—not as simple malfunction. This resonates deeply with [a]... thesis that cognitive 'failures' —including hallucinations, forgetting, and the emergence of psychopathology—are consequences of the same mechanisms that underlie cognitive power in dynamic, embodied contexts [4].

Specific Theoretical and Empirical Parallels

- The information theory account of cognitive control relates instability directly to limits on information processing, echoing [recent] formal modeling [5, 1].
- Embodied cognition approaches highlight emotion regulation failures and over-abstractness as routes to pathology, supporting the inevitability of instability without appropriate regulatory embedding [6, 3].
- Predictive coding models in psychopathology describe hallucination and delusion as failures of regulatory balancing, rather than simple technical errors [2].
- Recent work insists that 'noise' or variability is central to cognitive function, not a dispensable imperfection [4].

In summary, while... [the formal approach of this study]... articulates these dynamics with rigorous application of asymptotic limit theorems from information theory and nonequilibrium thermodynamics, the broader research landscape affirms the essential insight: that instability, regulation, and the ubiquity of potential failure are not design flaws, but deep, necessary features of real-world embodied cognition [1–4].

1. https://pmc.ncbi.nlm.nih.gov/articles/PMC4151034/2.
2. https://pmc.ncbi.nlm.nih.gov/articles/PMC5472128/3.
3. https://journals.sagepub.com/doi/10.5127/pr.0357144.
4. https://journals.sagepub.com/doi/full/10.1177/174569162412589515.
5. https://pmc.ncbi.nlm.nih.gov/articles/PMC7513233/6.
6. https://pmc.ncbi.nlm.nih.gov/articles/PMC5867482/.
7. https://www.frontiersin.org/journals/neuroscience/articles/10.3389/fnins.2025.1512724/full8.
8. https://www.nature.com/articles/s42003-023-05063-y9.
9. https://en.wikipedia.org/wiki/Modularity_of_mind10.
10. https://journalofcognition.org/articles/10.5334/joc.15111.

References

Anonymous: https://ai.stackexchange.com/questions/47924/do-mixture-of-experts-circumvent-the-limitations-imposed-by-the-no-free-lunch-th (2025)
Atlan, H., Cohen, I.: Immune information, selforganization, and meaning. Int. Immunol. **10**, 711–717 (1998)
Bennett, C.H.: The thermodynamics of computation. Int. J. Theor. Phys. **21**, 905940 (1982)
Cover, T., Thomas, J.: Elements of Information Theory, 2nd edn. Wiley, New York (2006)
Cyganowski, S., Kloeden, P., Ombach, J.: From Elementary Probability to Stochastic Differential Equations with MAPLE. Springer, New York (2002)
de Groot, S., Mazur, P.: Non-equilibrium Thermodynamics. Dover, New York (1984)
Dreyfus, H.: What Computers Can't Do: Of artificial reason. Harper and Row, New York (1972)
Dulberg, Z., Dubey, R., Betrwian, I., Cohen, J.: Having multiple selves helps learning agents explore and adapt in complex changings worlds. PNAS **120**, e2221180120 (2023). https://doi.org/10.1073/pnas.2221180120
Feynman, R.: The Feynman Lectures on Computation. Addison-Wesley (1996)
Feynman, R.: Statistical Mechanics: A Set of Lectures. Addison Wesley Longman (1998)
Fjelland, R.: Why general artificial intelligence will not be realized. Hum. Soc. Sci. Commun. (2020). https://doi.org/10.1057/s41599-020-0494-4
Khaluf, Y., Ferrante, E., Simoens, P., Huepe, C.: Scale invariance in natural and artificial collective systems: a review. Interface **14**, 20170662 (2017). https://doi.org/10.1098/rsif.2017.0662
Khinchin, A.: Mathematical Foundations of Statistical Mechanics. Dover, New York (1949)
Khinchin, A.: The Mathematical Foundations of Information Theory. Dover, New York (1957)
Kringlebach, M., Sanz Perl, Y., Deco, G.: The thermodynamics of mind. Trends Cogn. Sci. **28** (2024). https://doi.org/10.1016/j.tics.2024.03.009
Laidler, K.: Chemical Kinetics, 3rd edn. Harper and Row, New York (1987)
Landau, L., Lifshitz, E.: Statistical Physics, 3rd edn. Part 1. Elsevier, New York (2007)
Maturana, H., Varela, F.: Autopoiesis and Cognition: The Realization of the Living. Reidel, Boston (1980)
Nair, G., Fagnani, F., Zampieri, S., Evans, R.: Feedback control under data rate constraints: an overview. Proc. IEEE **95**, 108–138 (2007)
Ortega, P., Braun, D.: Thermodynamics as a theory of decision-making with information processing costs. Proc. Roy. Soc. A (2013). https://doi.org/10.1098/rspa.2012.0683

Osinga, F.: Science, Strategy and War: The Strategic Theory of John Boyd. Routledge, London (2007)
Perplexity AI: Comparing uses of the Lambert W function in physical phase transitions vs. cognition and consciousness models (2025)
Pettini, M.: Geometry and Topology in Hamiltonian Dynamics and Statistical Mechanics. Springer, New York (2007)
Roberts, K., Vallur, S.: The Lambert W Function, Laguerre Polynomials, and the Zeros of the QCD Partition Function (2014). https://arxiv.org/pdf/1307.1017
Saltini, A., Mishra, S., Reiner, P.: Nuclear Command, Control, and Communications: A Primer on Strategic Warning, Decision Support, and Adaptive Targeting Subsystems. Institute for Security and Technology (2025). https://securityandtechnology.org/
Shalev-Shwartz, S., Ben-David, S.: Understanding Machine Learning: From Theory to Algorithms. Cambridge University Press, New York (2014)
Stanley, H., Amaral, L., Gopikrishnan, P., Ivanov, P., Keitt, T., Plerou, V.: Scale invariance and universality: organizing principles in complex systems. Physics A **281**, 60–68 (2000)
Wallace, R., Fricchione, G.: Stress-induced failure of embodied cognition: a general model. BioSystems **239**, 105193 (2024). https://www.sciencedirect.com/science/article/abs/pii/S0303264724000789
Wallace, R.: Computational Psychiatry: A Systems Biology Approach to the Epigenetics of Mental Disorders. Springer, New York (2017)
Wallace, R.: How AI founders on adversarial landscapes of fog and friction. J. Defense Model. Simul. (2021). https://doi.org/10.1177/1548512920962227
Wallace, R.: Consciousness, Cognition and Crosstalk: The Evolutionary Exaptation of Nonergodic Groupoid Symmetrybreaking. Springer, New York (2022)
Wallace, R.: High frequency trading, artificial intelligence, and the instabilities of embodied cognition (2024). https://hal.science/hal-04763730v3
Wallace, R.: Detailed command vs. mission command: a cancer-stage model of institutional decision making. Stats **8**(2), 27 (2025). https://doi.org/10.3390/stats8020027
Wang, Z., Ying, Z., Bosy-Westphal, A., Zhang, J., Schautz, B., Later, W., Heymsfield, S., Muller, M.: Specific metabolic rates of major organs and tissues across adulthood: evaluation by mechanistic model of resting energy expenditure. Am. J. Clin. Neutr. **92**, 1369–1377 (2010)
Wang, J., Moniz, N.: Analysis of thermodynamic problems with the Lambert W function. Am. J. Phys. **87**, 752–757 (2017)
Wang, Y., Li, Y., Xu, C.: AI scaling: from up to down and out (2025). arXiv:2502.01677v1 [cs.LG] 2 Feb 2025
Weinstein, A.: Groupoids: unifying internal and external symmetry. Not. Am. Math. Assoc. **43**, 744–752 (1996)
Wikipedia: https://en.wikipedia.org/wiki/Born%E2%80%93Oppenheimer_approximation (2025)
Wikipedia: https://en.wikipedia.org/wiki/Lambert W function (2025)
Wolpert, D., MacReady, W.: No free lunch theorems for search. SFI-TR-02-010, Santa Fe Institute (1995)
Wolpert, D., MacReady, W.: No free lunch theorems and optimization. IEEE Trans. Evol. Comput. **1**, 67–82 (1997)

Chapter 6
Hypervigilance, Psychic Numbing and Bounded Rationality

...[C]omputational modelling is a promising avenue for advancing the understanding and treatment of PTSD.

—Skogbrott Birkeland and Sundnes (2024)

6.1 Introduction

Maturana and Varela (1980) argue that the living state is, effectively, cognitive at every scale and level of organization. Cognition, including matters of 'meaning'—semantics—as further characterized by Dretske (1994), is necessarily constrained by the asymptotic limit theorems of information theory. Atlan and Cohen (1998) find the essence of cognition, as represented by those theorems, to be the choice of response or responses to 'sensory' data that inform the system studied about itself and/or its environment. Choice reduces uncertainty in a formal manner, and the reduction of uncertainty implies the existence of an information source 'dual' to the cognitive process of interest. Nair et al. (2007), via the Data Rate Theorem, find that stability of an inherently unstable system requires pairing of cognition with regulation, i.e., imposition of control information at a rate greater than that system generates its own 'topological information', closely linking information and control theories.

Carrying matters further, fully *embodied* cognitive agents can be seen to act on, and be acted on by, embedding environments that may be less than benign or contain other agents having discordant intentions and abilities. Clausewitz characterizes something of this using terms like 'fog', 'friction', and 'will'. Sun Tzu explores similar discordances, from a different cultural perspective.

When performed as a systematic collective enterprise, such conflict imposes burdens at every scale and level of organized life.

© The Author(s), under exclusive license to Springer Nature Switzerland AG 2026

R. Wallace, *Bounded Rationality and Its Discontents*, Studies in Systems, Decision and Control 664, https://doi.org/10.1007/978-3-032-18423-8_6

Recent highly formal studies of such phenomena—a necessary synthesis across information and control theories (Wallace 2020, 2021, 2022a, b)—focused on a one-dimensional characterization of 'materiel/metabolic' supply rate, internal communication bandwidth, and the rate of sensory 'intelligence' available from the environment. An analog to this approach is reliance on the largest vector in the familiar Principal Component Analysis of a multivariate system. Here, we attempt a more direct and less draconian extension of the formalism than was provided by Wallace (2021).

Here, we begin to address the impact of increasing Clausewitzian fog on an otherwise stable cognition rate in such a multidimensional, multidomain milieu, the onset of what might be called 'stochastic anxiety'. The constant triggering of exacerbated 'false alarm' responses in that fog can carry severe consequences if such triggering is, in fact, part of a larger adversarial strategy—or selection pressure—aimed at systematic desensitization or exhaustion of an overconfident opponent.

Other applications are to the study of culture-bound anxiety/depression disorders expressed at the individual rather than institutional or machine levels (Wallace 2017, Ch. 3; Chiu et al. 2024; Ehlers and Clark 2000; Korem et al. 2024). The pathologies of cognitive physiological subsystems may also benefit from this work, for example understanding autoimmune attack and immune exhaustion, and dysfunctions of the HPA axis.

It should be clear that critical-purpose machine cognition 'artificial intelligence' systems will not be spared recognizably similar failings (Wallace 2025a, b).

Conversion of the asymptotic limit theorems of information and control theories to statistical tools for the analysis of data related to cognition and its dysfunctions may seem, at first, like a disorienting terra incognita. Recall that the Central Limit Theorem and Friends led to the current spectrum of regression models, t-tests, and the like.

While statistical analysis is not, of itself, science, the exploration of experimental and observational data using statistical methods can sometimes aid in the conduct of science.

The One-Dimensional Approximation

There are at least three resource streams necessary for survival of real-world, real-time systems in-and-of an embedding environment:

1. The rate at which elements of the system can communicate internally with each other, measured by a channel capacity $\mathscr{C}$.
2. The rate at which 'sensory information' is available from the environment, measured as another channel capacity $\mathscr{H}$.
3. The rate at which 'material/materiel' resources can be provided, $\mathscr{M}$. For an organism, this could be the rate at which metabolic free energy is provided.

These quantities interact and cross-correlate, generating a 3×3 matrix $\mathbf{Z}$. An $n \times n$ square matrix has n scalar measures r_i invariant under similarity transformations, as determined by

$$p(\gamma) = \det[\mathbf{Z} - \gamma \mathbf{I}] = \sum_{k=0}^{n} (-1)^k r_k \gamma^{n-k} \tag{6.1}$$

I is the n-dimensional identity matrix, 'det' the determinant, and γ a real-valued parameter. The first is the matrix trace, and the last the determinant. We assume it possible to build a scalar resource rate measure from the r_k that captures much or most of the underlying dynamics. This may not be true.

An elementary one-dimensional index might be $Z = \mathscr{C} \times \mathscr{H} \times \mathscr{M}$, but circumstances are not likely to be that simple.

A general first-order model assumes that a single scalar index $Z(r_1, \ldots, r_n)$ can be constructed that sufficiently represents information and material resource rates. Again, this argument is analogous to principal component analysis based on a correlation matrix. And again, Wallace (2021) outlines some of the general difficulties inherent to more than a single such index.

Here, we will explore a simplified formalism that extends the one-dimensional analysis in a consistent manner.

Rate Distortion Control Theory

'Embodied' formalism emerges via a bounded rationality Rate Distortion Control Theory (RDCT) based on the Rate Distortion Theorem (RDT) (Cover and Thomas 2006; Khinchin 1957). The RDT addresses information transmission under conditions of noise over a given channel. For that channel and a given scalar measure of average distortion D between what is sent and what is received, there will be a minimum necessary channel capacity $R(D)$.

The theorem asks, effectively, what is the 'best' channel for transmission of a message with the least possible average distortion. $R(D)$ can be defined for nonergodic information sources via a limit argument based on the ergodic decomposition of a nonergodic source into a 'sum' of ergodic sources. The RDT can be reconfigured in a control theory context, if we envision a system's topological information from the Data Rate Theorem (DRT) (Nair et al. 2007) as another form of noise, adding to the average distortion D. See Fig. 6.1.

The DRT states that, for stabilizing an inherently unstable system, control information must be imposed at a rate greater than the system counters with its generation of 'topological information'. Envision a vehicle driven at high speed on a twisting, pot-holed road. The driver must steer, brake, and shift gears at a rate greater than the road imposes its topology, at the chosen speed.

The punctuation implied by the DRT occurs in RDCT if there is a maximum distortion average beyond which control fails.

Embodied cognition, characterized by continuous interaction with a real world having its own intents and capabilities—as indexed by the distortion between what is wanted and what is observed—instantiates the 'bounded rationality' of Ortega and Braun (2013). Cognition without embodied interaction, as Dreyfus (1972), Fjelland (2020), and others, have noted, implies a distinct absence of rationality.

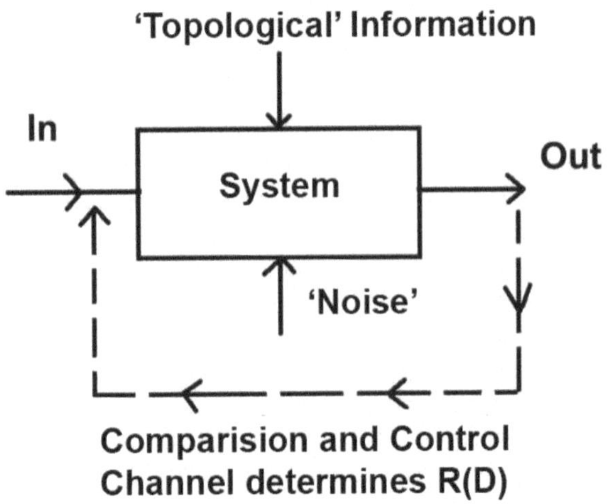

Fig. 6.1 RDCT. An inherently unstable cognitive system's 'topological information', according to the Data Rate Theorem, is added to the effects of 'noise' to determine the minimum channel capacity $R(D)$ needed to assure 'transmission' of system intent on the embedding environment with average distortion D. The punctuation implied by the DRT occurs if there is a critical maximum average distortion beyond which control fails

6.2 Dynamics of Cognition Under Uncertainty

The Basic Idea

Consider an ensemble of high probability developmental trajectories available to the system, say Y_j, $j = 1, 2, \ldots$. We can associate each with a Rate Distortion Function-defined minimum necessary channel capacity R_j for a particular maximum average scalar distortion D_j. Then, given a basic underlying probability model having distribution function $\rho(x)$, we can define a pseudoprobability for a meaningful 'message' Y_j sent into the system of Fig. 6.1 as

$$dP = \frac{\rho(R/g(Z))dR}{\int_0^\infty \rho(R/g(Z))dR} = \frac{\rho(R_j/g(Z))dR}{g(Z)} \tag{6.2}$$

since $\rho(x)$ is a probability density function that integrates to one on the interval $x = 0 \rightarrow \infty$.

$g(Z)$ is taken as a 'temperature' analog that—here—must be further calculated from first principles, as it depends on the resource rate Z.

$g(Z)$ is, in a sense, inverse to the 'rationality parameter' that controls the trade-off between utility and information processing cost in theories of bounded rationality (Genewein et al. 2015).

For simplicity, we treat the continuous case. Under discrete or finite-state circumstances, the integral is a sum and the partition function denominator becomes a general function $h(g(Z))$.

Again, Y_j is some particular system trajectory, while the integral is over all possible 'high probability' paths.

We are implicitly imposing some version of the Shannon-McMillan Source Coding Theorem (Cover and Thomas 2006): the set of possible system trajectories can be divided into two sets: First, a very large set of measure zero—vanishingly low probability—not consistent with the underlying grammar and syntax of some basic information source, and, second, a much smaller high probability set that is consistent (Khinchin 1957).

Recall that R_j is the RDT channel capacity keeping average distortion less than a given limit D_j for message Y_j. $g(Z)$ is a temperature analog depending on the scalar resource rate Z. $\rho(x)$, for physical systems, is usually taken as the Boltzmann distribution: $\rho(x) = \exp[-x]$. In general, for cognitive phenomena—from the living and institutional to the machine and composite—it will be necessary to move beyond analogs with physical systems. That is, it becomes necessary explore the influence of a variety of different probability distributions, including those with 'fat tails' (Derman et al. 2016; Taleb 2020), on the dynamics of cognition/regulation stability. Here, however, for a first-cut higher dimensional model, we will focus on the Boltzmann Distribution.

Methodology from statistical physics (Landau and Lifshitz 2007) identifies the denominator of Eq. (6.2) as a partition function, giving an iterated free energy-analog F as

$$\rho(F/g(Z)) \equiv \int_0^\infty \rho(R/g(Z))dR = g(Z)$$
$$g(Z) = \text{RootOf}\,(X - \rho(F/X)) \tag{6.3}$$

where 'RootOf' means setting the expression to zero and solving for X and ρ is an appropriate 'free energy' distribution that must be determined case-by-case for cognitive, as opposed to simple physical, phenomena. For physical phenomena, ρ is often taken as the Boltzmann distribution.

From chemical physics (Laidler 1987), we can define a cognition rate for the system as the reaction rate analog

$$L \equiv \frac{\int_{R_0}^\infty \rho(R/g(Z))dR}{\int_0^\infty \rho(R/g(Z))dR} = \frac{1}{g(Z)} \int_{R_0}^\infty \rho(R/g(Z))dR \tag{6.4}$$

R_0 is the minimum channel capacity needed to keep the average distortion below a critical value D_0 for the full system of Fig. 6.1.

For the Boltzmann, second-order Erlang, Rayleigh, and Chi-Square Distributions, direct calculation finds expressions for g and L as

$$\rho(x) = e^{-x},\ g = \frac{-F}{W(m,\ -F)},\ L = \exp[-R_0/g]$$
$$\rho(x) = x\,e^{-x},\ g = -\frac{F}{2\,W\left(m,\ -\frac{\sqrt{F}}{2}\right)},\ L = e^{-\frac{R_0}{g}}\left(1 + \frac{R_0}{g}\right)$$

$$
\rho(x) = x\,\mathrm{e}^{-\frac{x^2}{2}},\ g = -\frac{F}{\sqrt{-2\,W\left(m, -\frac{F}{2}\right)}},\ L = \mathrm{e}^{-\frac{R_0^2}{2g^2}}
$$

$$
\rho(k, x) = \frac{x^{\frac{k}{2}-1}\mathrm{e}^{-\frac{x}{2}}}{2^{\frac{k}{2}}\Gamma\left(\frac{k}{2}\right)},\ g = -\frac{F}{kW\left(m, -\frac{\left(F 2^{\frac{k}{2}}\Gamma\left(\frac{k}{2}\right)\right)^{\frac{2}{k}}}{k}\right)}
$$

$$
L = \frac{\Gamma\left(\frac{k}{2}, \frac{R_0}{2g}\right)}{2^{\frac{k}{2}}\Gamma\left(\frac{k}{2}\right)\left(\frac{1}{2}\right)^{\frac{k}{2}}} \tag{6.5}
$$

$W(m, x)$ is the m-branch of the Lambert W-function that satisfies the relation

$$
W(m, x)\exp[W(m, x)] = x
$$

where m is an integer.

We will show below how monotonic and nonmonotonic distributions express markedly different cognitive dynamics under arousal.

Several matters regarding the W-function are important:

- $W(m, x)$ has real-valued range only for $m = 0, -1$.
- The $m = 0$ branch is real-valued for $x > -\exp[-1]$, but the $m = -1$ branch only for $-\exp[-1] \leq x \leq 0$.
- These range differences will prove crucial, allowing characterization of the $m = 0$ branch as an 'on' mode and the $m = -1$ branch as 'off'.
- The occurrence of the Lambert W-function in the calculations of g for so many distributions implies the phenomena studied here may be quite general.

In regard to the last point, the Lambert W-function typically arises when the variable appears both as a factor and in an exponent. In the context of probability distributions, this happens when one imposes an implicit normalization or self-consistency condition—such as equating the pdf at some transformed argument to a constant or function of itself. Consequently, W enables explicit inversion of these functional relations, giving analytical expressions for normalization constants or transformed variables that would otherwise require iterative numerical methods.

In short, solving $\rho(F/g) = g$ for g often reduces to a standard Lambert W inversion problem—turning an implicit transcendental relationship into a closed explicit form that preserves analytic tractability in models of exponential, gamma, or heavy-tailed distributions.

The basic underlying probability model of Eq. (6.2)—via the partition function argument of Eq. (6.3)—determines system dynamics. The argument may perhaps seem somewhat counterintuitive, as the basic assumption is that information is not an 'entropy', but a form of free energy, in spite of the algebraic form of the expression for Shannon uncertainty in a stationary ergodic system.

On this matter, Feynman (2000) identifies 'information' as a free energy, using Bennett's (1982) clever ideal machine that turns a message directly into work. We use this perspective to impose an iterated, rather than a simple direct, free energy construction.

F, as a free energy, is subject to symmetry-breaking transitions as g varies (Pettini 2007). These symmetry changes, however, are not like those associated with physical phase transitions as represented by standard group algebras. Such 'cognitive' symmetry changes represent transitions from playing one kind of 'game' to some other. For example, an entity may engage in foraging behaviors that trigger a predatory attack by another agent. Then the game is changed from 'foraging' to 'escape'.

Thus 'cognitive phase transitions' involve shifts between equivalence classes of high probability developmental/behavioral pathways that are represented as groupoids, a generalization of the group concept such that a product is not necessarily defined for every possible element pair, although multiple products with multiple identity elements are defined (Brown 1987; Weinstein 1996).

Noise Dynamics

Dynamics emerge via a first-order Onsager approximation akin to that of nonequilibrium thermodynamics (de Groot and Mazur 1984) in the gradient of an entropy measure S constructed from the 'iterated free energy' F of Eq. (6.3) by means of a Legendre transform familiar from the Black-Scholes approximation (Landau and Lifshitz 2007; Pettini 2007).

We are, however, now concerned with a multidimensional system having $p \leq n$ independent Z-components.

First-order dynamics follow from an Onsager-like approximation

$$
\begin{aligned}
S &\equiv -F + \sum_j Z_j \partial F / \partial Z_j \\
dZ_j/dt &\propto \partial S / \partial Z_j = \sum_i Z_i \partial^2 F / \partial Z_i \partial Z_j
\end{aligned}
\tag{6.6}
$$

where, for algebraic convenience, we set the associated 'diffusion coefficient(s)' equal to one.

'Noise' can now be imposed via a set of stochastic differential equations

$$
dZ_t^j = \left(\partial S / \partial Z_t^j \right) dt + \sigma^j Z_t^j dB_t^j \tag{6.7}
$$

where the dB_t^j represent Brownian white noise and the last terms express 'ordinary' volatility.

We set all $\sigma_j \equiv \sigma$, and *require stability in second order for survival*—stability in variance—so that, defining $Q \equiv \sum_i Z_i^2$, $\langle dQ_t \rangle = 0$ under the Ito Chain Rule. The resulting expression is an elaborate differential equation defining F in terms of the Z_j and σ.

For $j = 1, 2$, that equation becomes

$$
\begin{aligned}
&2\left(Z_1\left(\frac{\partial^2}{\partial Z_1^2}F(Z_1, Z_2)\right) + Z_2\left(\frac{\partial^2}{\partial Z_1 \partial Z_2}F(Z_1, Z_2)\right)\right)Z_1 \\
&\quad + 2\left(Z_1\left(\frac{\partial^2}{\partial Z_1 \partial Z_2}F(Z_1, Z_2)\right) + Z_2\left(\frac{\partial^2}{\partial Z_2^2}F(Z_1, Z_2)\right)\right)Z_2 \\
&\quad + \sigma^2\left(Z_1^2 + Z_2^2\right) = 0
\end{aligned}
\tag{6.8}
$$

having the surprisingly simple general solution

$$
F = -\frac{\sigma^2}{4}\left(Z_1^2 + Z_2^2\right) + Z_1 f_1\left(\frac{Z_2}{Z_1}\right) + f_2\left(\frac{Z_2}{Z_1}\right)
\tag{6.9}
$$

We set $f_2(x) \equiv C_2$ and $f_1(x) = C_1\sqrt{x}$, leading to the geometric average of Z_1 and Z_2. Taking $f_1 = (C_1/2)(x + 1)$ provides the arithmetic average.

More explicitly,

$$
F = -\frac{\sigma^2}{4}\left(Z_1^2 + Z_2^2\right) + C_1\sqrt{Z_1 Z_2} + C_2
\tag{6.10}
$$

The different Z_j postulated here may not, in fact, simply be independent components of an 'affordance' vector derived from Eq. (6.1). We can, in Eq. (6.10), assign Z_2 to an affordance, but Z_1 to an adrenaline-driving selection pressure, e.g., the influence of a cognitive adversary or an adverse environmental circumstance under conditions of fog, friction, and deadly intent or blind effect.

For the monotonic Boltzmann case, invoking Eq. (6.5) in Fig. 6.2a, we calculate two cognition rates as functions of Z_2, setting $Z_1 = 4,\ C_1 = -1,\ C_2 = 3$, with $R_0 = 1$, under two values of 'noise', $\sigma = 0,\ 0.7$ as indicated.

Note, however, the two solution forms corresponding to $m = 0, -1$ in the Lambert W-function defining g. The $m = 0$ form we might characterize as 'on', the $m = -1$ form as 'off'.

The 'on' cognition rates L versus arousal Z_2 follow a Yerkes-Dodson effect pattern for an 'easy' problem (e.g., Wallace and Fricchione 2024). However, increasing the 'noise' σ triggers response at a progressively lower arousal threshold. This will prove important.

For the second-order Erlang Distribution, Fig. 6.2b, taking the same parameter values as for the Boltzmann case, matters are far different. Here, we see something more like the inverted-U of the 'hard' problem under the Yerkes-Dodson effect. Again, there are 'on' and 'off' states, and again rising σ triggers response at a lower arousal threshold.

In Fig. 6.3, for the Boltzmann model, we allow both Z_1 and Z_2 to vary, keeping other parameters the same as in Fig. 6.2. Again, and in general, rising 'noise' σ decreases the trigger threshold for cognitive response, driving the system to higher cognitive effort at a given level of arousal. Again, the lower, darker rims represent the

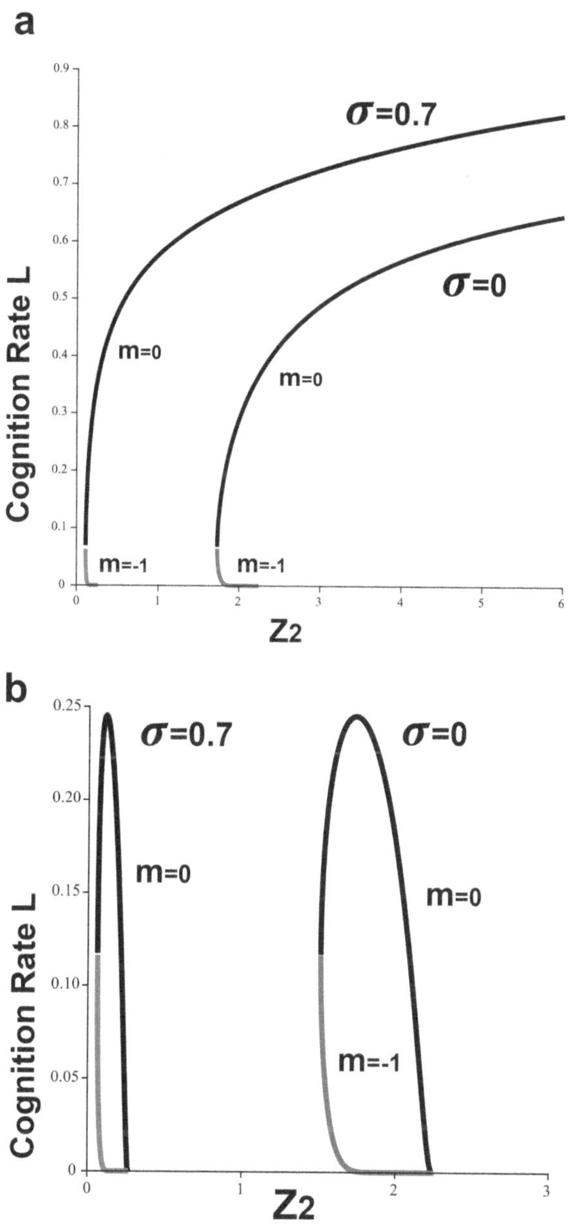

Fig. 6.2 **a** For the monotonic Boltzmann model of Eq. (6.5), we fix $Z_1 = 4$ and calculate two cognition rates as functions of Z_2, setting $C_1 = -1,\ C_2 = 3$, with $R_0 = 1$, under two values of 'noise', $\sigma = 0,\ 0.7$ as indicated. There are, for each, two solutions corresponding to the $m = 0,\ -1$ branches of the Lambert W-function. We may well call the $m = 0$ branches 'on' and the $m = -1$ branches 'off'. This resembles the 'easy' form of the Yerkes-Dodson effect. **b** The nonmonotonic second-order Erlang model, for the same parameter values as the Boltzmann example, produces a qualitatively different pattern, more akin to the inverse-U of 'difficult' Yerkes-Dodson problems, although with the same 'on' and 'off' dichotomy

$m = -1$ 'off' branches, and the lighter shells the $m = 0$ 'on' condition. A sufficient rate of 'on' false alarms can enervate and exhaust response practice and policy at both institutional and individual levels of organization, driving the system into the 'off' mode.

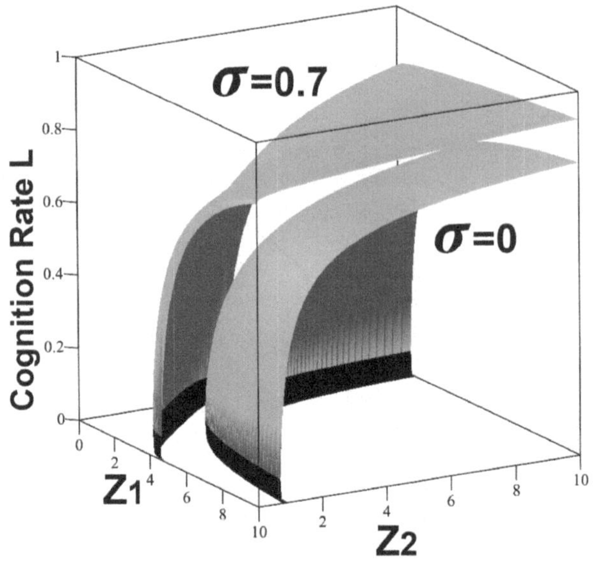

Fig. 6.3 Extension of Fig. 6.2a, letting Z_1 vary at the same settings. This is essentially a rotation about the vertical axis. Rising 'noise' σ triggers greater cognitive sensitivity, ultimately causing enervated response to 'false alarms'. The lower, darker rims represent the $m = -1$ 'off' branches, and the lighter shells the $m = 0$ 'on' condition. A high rate of 'on' false alarms can exhaust the capability of response, driving the system permanently into the 'off' mode. The second-order Erlang model of Fig. 6.2b similarly rotates about the vertical axis as Z_1 varies

The second-order Erlang model calculation simply rotates Fig. 6.2b about a vertical axis.

For the Boltzmann model, the expression for g in Eq. (6.5), via the Lambert W-function of orders $m = 0, -1$ becomes complex-valued when

$$F = -\frac{\sigma^2 \left(Z_1^2 + Z_2^2\right)}{4} + C_1\sqrt{Z_2 Z_1} + C_2 = e^{-1} \tag{6.11}$$

For the second-order Erlang model, this condition is $\sqrt{F}/2 = e^{-1}$.

In Fig. 6.4, we apply this condition to the system of Fig. 6.3. Points above the surface give real-valued outcomes. The second-order Erlang case is quite similar.

Note that, in Fig. 6.4, from Eq. (6.11), the 'stochastic anxiety' plane for the Boltzmann model, where $Z_2 = 0$, is defined by

$$\sigma = \frac{2\sqrt{(C_2 - \exp[-1])}}{Z_1} \tag{6.12}$$

That is, the critical σ that drives the system to permanent alert declines with rising 'adversarial' pressure Z_1.

The expression for the Erlang model is analogous, i.e., $\sigma \propto 1/Z_1$.

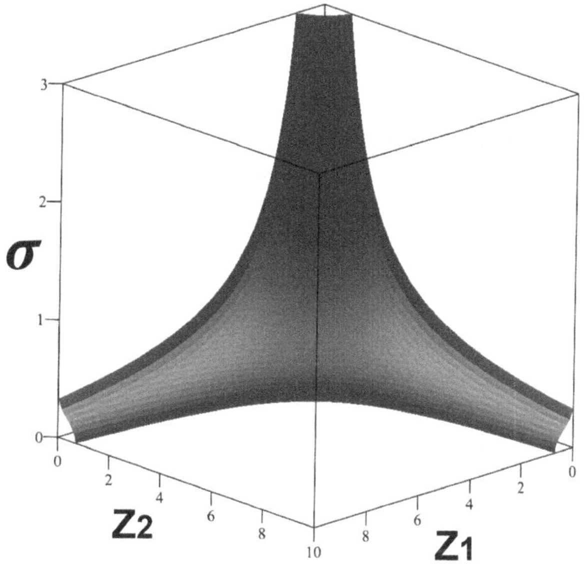

Fig. 6.4 Stability criterion of Eq. (6.11) for the system of Fig. 6.3. Points above the surface give real valued outcomes. The pattern for the second-order Erlang model is quite similar

Given such permanent alert, the $m = 0$ branch of cognition rate would seem to characterize lock-in to a hypervigilant state, while transition to the $m = -1$ mode would seem to indicate onset of classic psychic numbing (Lifton 1982).

For 'fat-tailed' distributions (Derman et al. 2016; Taleb 2020), without well-defined mean or variance, matters become dicey indeed. First, it is not entirely clear that $\langle dQ_t \rangle$ for $Q = Z_1^2 + Z_2^2$ can be well-defined under such conditions. However, proceeding blindly for the Lomax, Levy, and Cauchy distributions—what physicists call 'letting the notation carry us through'—does produce pathological versions of Fig. 6.2. These do indeed split into two cognition rate equivalence classes, the 'on' and 'off' modes, that are likewise driven toward the origin by increasing 'noise' σ, but the response patterns are highly distorted. As in financial engineering, systems dominated by pathological distributions seem destined to experience profound dysfunction. There appear to be significant mechanisms hidden some layers further within these observations.

6.3 The Synergism of Plagues

The arguments of Eqs. (6.6)–(6.10) can 'easily' be extended to higher dimensions, almost by induction. We consider a 'two-front war', supposing that a base system, represented by the resource rate Z_3 engages two adversaries, represented as Z_1, Z_2, at the same time.

Examples might include an organism's immune response facing two pathogens simultaneously, an island nation with limited resources triggering a two-front war against better-situated opponents, or a worker experiencing both job and family problems. An emerging realm of concern might involve high-level AI systems tasked with management of critical infrastructure being confronted by discordant priorities.

The canonical picture is of institutions, communities, individuals, sophisticated AIs, or their varied composites and cognitive subcomponents, torn by Scylla-and-Charybdis 'wickedly hard' problems.

Perhaps the simplest 'realistic' extension of the previous calculation for F leads, somewhat surprisingly, to

$$F = -\frac{\sigma^2}{4}\left(Z_1^2 + Z_2^2 + Z_3^2\right) + C_1 Z_1 + C_2 Z_2 + C_3 Z_3 + C_4 \qquad (6.13)$$

Derivation is left as a not entirely trivial exercise.

Assuming a Boltzmann Distribution, according to the first line of Eq. (6.5), gives Fig. 6.5. Here, $Z_1 = Z_2 = 1$, $C_1 + C_2 + C_4 = 3$, $C_3 = -1$ and $\sigma = 0, 1.4$. The outcome is similar to Fig. 6.2a.

Results for the nonmonotonic second-order Erlang Distribution are again left as an exercise.

6.4 Yet Another Idiotypic Conundrum

Recent work by Korem et al. (2024) explores rapid transitions between 'neutral' to 'negative valence' states in subjects with probable PTSD in comparison with trauma-exposed 'normal' controls, as shown in Fig. 6.6, adapted from Fig. 2 of their paper.

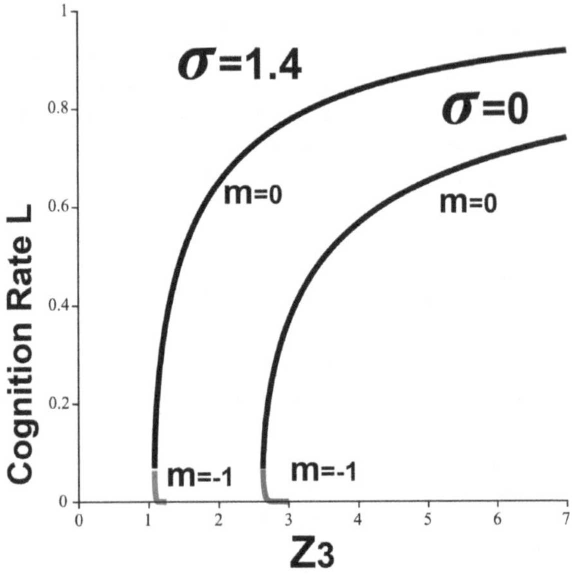

Fig. 6.5 Under a Boltzmann Distribution, assuming Eq. (6.13), the pattern for a two-adversary challenge is similar to Fig. 6.2a. Again, increasing 'noise' triggers premature response in the context of equivalence class phase transitions

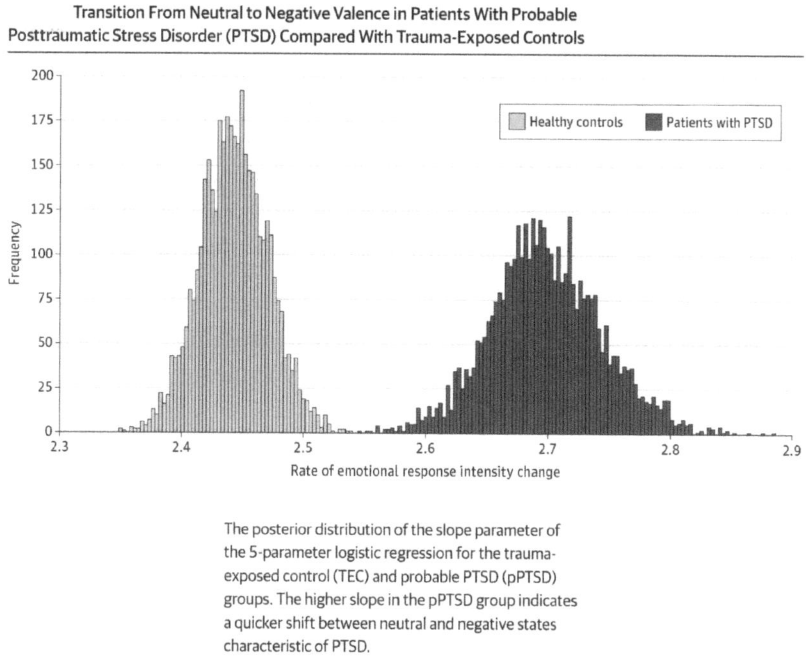

Fig. 6.6 Adaptation of Fig. 2 of Korem et al. (2024). Individuals with PTSD undergo rapid transitions between neutral and negative emotional states, as intensified by the severity of emotional numbing symptoms

Korem et al. (2024) conclude

> ...[I]ndividuals with PTSD undergo rapid transitions between neutral and negative emotional states, a phenomenon intensified by the severity of emotional numbing symptoms. Therapeutic interventions aimed at moderating these swift emotional transitions could potentially alleviate PTSD symptoms.

In the terminology of this chapter, Korem et al. observe rapid oscillation between the ‘on’ and ‘off’ states characterized by Figs. 6.2, 6.3, 6.4 and 6.5 that are associated with the expression of underlying probability distributions across a broad spectrum of cognitive phenomena. Many such distributions are inherently constrained by the dynamics of Eq. (6.3), leading to the $m = 0, -1$ (or, for Lomax and like distributions, corresponding $\pm$) states defined by inherent ‘symmetries,’ here represented by the Lambert W-function.

Hypersensitivity and/or pathological desensitization under persistent ‘false alarm’ is also strongly expressed in dysfunctions of the immune and HPA axis systems, as described in the Appendix by the Perplexity AI system.

The narrative, and the realm of application of the models, can be expanded to the parallel and intensely cognitive phenomena of protein folding and gene expression.

The identification of discordant dynamics associated with the $m = 0, -1$ states defined by the Lambert W-function implies deeper matters: the existence of (at least two) equivalence classes of response patterns, with highly punctuated transformations possible between or among them. As frequently described in the Wallace references, this is yet another example of the 'groupoid symmetry-breaking phase transitions' inevitably associated with most if not all cognitive phenomena.

We have not, however, yet provided formal models for explicit mechanisms by which transitions between different states or modes are triggered. For physical systems, changes in ordinary temperature often drive group symmetry-breaking phase changes. Here, one will apparently need to impose another full iteration of theory to provide possible channels for shift within some small set of available modes. This seems, on the surface, deceptively straightforward.

We begin by exploring a 'simple' two-state systems using the previous methods, defining free energy and 'reaction rate' analogs, and solving for 'temperature' g:

$$
\begin{aligned}
&\exp[-F/g] = \exp[-(R_0+\delta)/g] + \exp[-(R_0-\delta)/g]\\
&L \equiv \frac{\exp[-(R_0+\delta)/g]}{\exp[-(R_0+\delta)/g] + \exp[-(R_0-\delta)/g]}\\
&= \frac{1}{2}\mathrm{e}^{-\frac{\delta}{g}}\operatorname{sech}\left(\frac{\delta}{g}\right)\\
&g_{\pm} \approx -\frac{F - R_0 \pm \sqrt{(F-R_0)^2 - 2\ln(2)\,\delta^2}}{2\ln(2)}
\end{aligned}
\tag{6.14}
$$

assuming, for the 'energy difference' between the states, $\delta \ll R_0$.

First, there are—yet again—two forms of 'temperature', represented as $g_{\pm}$.

Second, critical transitions occur in both expressions for g when

$$
(F - R_0)^2 = \ln(4)\delta^2 \tag{6.15}
$$

analogous to the previous $m = 0, -1$ dichotomies. Again, there are two equivalence classes of solutions, corresponding to the $g_{\pm}$.

For nonmonotonic probability distributions on $[0, \infty)$—Erlang, Rayleigh, Gamma, etc.—the approximation can be written

$$
\begin{aligned}
&\rho\left(\frac{F}{g}\right) \approx 2\rho\left(\frac{R_0}{g}\right) + \frac{D^{(2)}(\rho)\left(\frac{R_0}{g}\right)\delta^2}{g^2}\\
&F \approx \mathit{RootOf}\left(-\rho(Z)\,g^2 + 2\rho\left(\frac{R_0}{g}\right)g^2 + D^{(2)}(\rho)\left(\frac{R_0}{g}\right)\delta^2\right)g
\end{aligned}
\tag{6.16}
$$

The number of possible solutions for F is then determined by the relation

$$\rho(Z) = \frac{2\rho\left(\frac{R_0}{g}\right) g^2 + D^{(2)}(\rho)\left(\frac{R_0}{g}\right) \delta^2}{g^2} \equiv K \tag{6.17}$$

For nonmonotonic distributions, if $0 < K < \rho_{\max}$, there will be at least two solutions, and thus two or more equivalence classes defining a new groupoid symmetry-breaking phase transition.

At the other end of approximation is the continuous relation

$$\rho(F/g) = \int_0^\infty \rho(R/g) dR = g \tag{6.18}$$

assuming $\rho(x)$ is a probability distribution on $[0, \infty)$.

- If $\rho(x)$ monotonically decreases from zero like the Boltzmann Distribution, $d\rho(F/g)/dg > 0$, so that $\rho(F/g)$ is a sigmoid function of g. If Eq. (6.18) has a nonzero solution, it is almost certain to have two solutions, as the straight line for increasing g cuts through the hump of the S-shaped curve.
- If $\rho(x)$ is nonmonotonic, then an analog to the considerations of Eq. (6.17) applies, and there will be at least two solutions for F. In addition, double solutions for g may also sometimes be possible.

Without the need to iterate further, we have uncovered distressing analogs to Jerne's (1974) *Idiotypic Network* problem: there is no 'natural' end to this recursive nesting for the study of PTSD-like thrashing between 'on' and 'off' modes, or progression across some continuous spectrum of symptoms.

There is quite an extensive literature on the intractability of recursion/idiotypic network problems. See, for example, Bartolucci et al. (2016), who attempt detailed statistical mechanical methods.

While mathematical models can provide some characterization and scaling results for Jerne's idiotypic network (and their analogs in cognition and AI), the infinite-layer problem remains analytically, and possibly fundamentally, unsolved—exhibiting fragmentation, multi-attractor chaos, and phase transitions rather than convergence.

As De Boer and Hogeweg (1989) put it some time ago,

> ...[O]ur results cast doubt of the functional role of a profound idiotypic network. The network fails to account for proliferation regulation, and if it accounts for memory phenomena, it 'explodes' upon the first encounter with antigen due to extensive percolation...[contradicting] our intuitive expectations about the functioning of this simple (fundamental) network.

Again, Terra Incognita.

6.5 Toward Simplification: Equipartition

Equation (6.14) implicitly assumes a given R_0 and a 'small' δ separating two different possible pathological 'states'. Under conditions of sufficient 'churning', there may be very many possible pathological states that share what might be called the same underlying energy landscape. In complex physical systems, particularly at some given temperature, free energy tends to be equally distributed among possible excitation modes, for example different kinds of vibration under the influence of a stochastic excitation. The phenomenon ranges from molecular structures with very many possible conformations that converge on equally-distributed energy states to bridges vibrating in a high, gusty wind.

Here, under some foundational probability distribution ρ, such equipartition manifests as

$$
\begin{aligned}
&\rho\left(\frac{F}{\langle g\rangle}\right) = \sum_{j=1}^{n} \rho(R_j/g_j) \equiv n\ \rho(X) \\
&R_j/g_j \equiv X\ \forall j \\
&\langle g\rangle \equiv \frac{\langle R\rangle}{X} \\
&\rho\left(\frac{F}{\langle g\rangle}\right) = \rho\left(\frac{FX}{\langle R\rangle}\right) = n\ \rho(X)
\end{aligned}
\tag{6.19}
$$

where n represents the number of possible—here pathological—equipartition modes. $\langle g\rangle$ is taken as an average temperature-analog and $\langle R\rangle$ an average channel capacity across all possible modes.

The dynamic equations are again—as a simplistic first-cut approximation—abducted from nonequilibrium thermodynamics:

$$
\begin{aligned}
&S \equiv -F(X) + XdF/dX \\
&dX/dt \propto dS/dX = Xd^2F/dX^2 \\
&dX_t = (X_t d^2F/dX^2)dt + \sigma X_t dB_t
\end{aligned}
\tag{6.20}
$$

where dB_t represents Brownian noise and the last term in the third relation implements ordinary volatility.

Requiring stability in second order under the Ito Chain Rule, so that $\langle dX_t^2\rangle = 0$,

$$
\begin{aligned}
&F = -\frac{1}{4}\sigma^2 X^2 + C_1 X + C_2 \\
&\langle R\rangle = \frac{F(\sigma,\ X)X}{\rho^{-1}(n\ \rho(X))}
\end{aligned}
$$

$$\langle R \rangle = -\frac{\left(-\frac{\sigma^2}{4}X^2 + C_1 X + C_2\right) X}{\ln(n) - X} \tag{6.21}$$

where the last expression is for the Boltzmann Distribution.

Under that distribution, Fig. 6.7 shows $\langle R \rangle(X)$ for $n = 2$, $C_1 = 1/10$, $C_2 = 2$ and $\sigma = 0,\ 1.4$.

After a 'hallucination' mode, the noise-free system converges.

From the second expression in Eq. (6.21)—for any distribution and at any given X—sufficient noise crashes out the system, driving F, and hence $\langle R \rangle$, to 0. This, of course, makes the average distortion D very large indeed since $R(D)$ is always convex in D:

$$F(\sigma_{Crit}, X) = 0$$
$$\sigma_{Crit} = \frac{2\sqrt{C_1 X + C_2}}{X} \tag{6.22}$$

A contrasting approach is via the required stabilization of a 'perceived signal' as represented by the Weber-Fechner Law (2025):

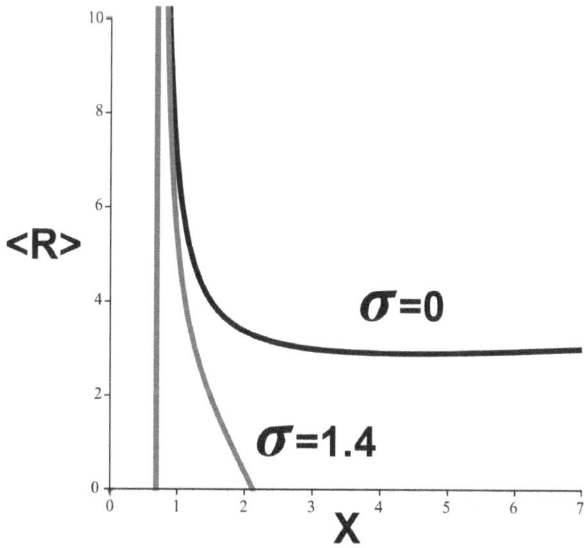

Fig. 6.7 The impact of 'noise' under equipartion in the expression of different modes of pathology. Here, under the Boltzmann Distribution, $\sigma = 0,\ 1.4$, with $n = 2$, $C_1 = 1/10$, $C_2 = 2$. At any given X, and, following Eq. (6.22), any distribution, sufficiently large σ drives $\langle R \rangle$ to zero, crashing the system

$$P \approx \alpha \ln\left(\frac{X}{X_0}\right) \tag{6.23}$$

Then, setting $\langle dP_t \rangle = 0$ under the Ito Chain Rule, somewhat surprisingly,

$$F = \frac{1}{4}\sigma^2 X^2 + C_1 X + C_2 \tag{6.24}$$

with $\langle R \rangle$ still defined as in Eq. (6.21).

Under Weber-Fechner, given the same the boundary conditions as the $\langle dX_t^2 \rangle = 0$ example, a different pattern emerges, shown in Fig. 6.8. Again, $C_1 = 1/10,\ C_2 = 2$, again with $\sigma = 0,\ 1.4$. The $\sigma = 0$ system, as expected, after a brief 'hallucination' mode, converges to a stable state in $\langle R \rangle(X)$. The 'noisy' system, by contrast, explodes, requiring massive levels of bandwidth for stability.

For the Stevens Law of psychophysics

$$P \approx kX^a \tag{6.25}$$

with, typically, $0.5 \leq a \leq 3.5$.

The nonequilibrium steady state Ito Chain Rule calculation for $\langle dP_t \rangle = 0$ based on Eq. (6.25) gives

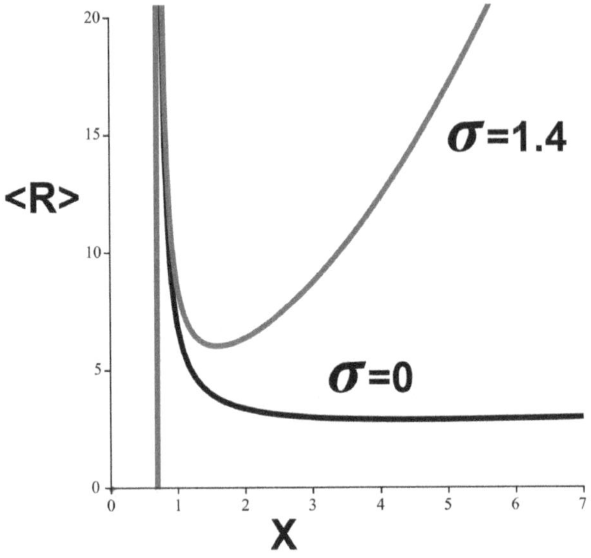

Fig. 6.8 Assuming Eq. (6.24) under the Weber-Fechner condition $\langle dP_t \rangle = 0$ and the Boltzmann Distribution, again with $C_1 = 1/10,\ C_2 = 2$. Here, nonzero 'noise' leads to an ultimately explosive bandwidth demand for maintenance of stability

$$F = -\frac{\sigma^2}{4}(a-1)X^2 + C_1X + C_2 \tag{6.26}$$

generating patterns ranging from Fig. 6.7 to Fig. 6.8. MacKay (1963), in fact, derives a version of the Stevens Law from the Weber-Fechner Law.

Under equipartition, the effects of 'noise' can be far more subtle than described by the mechanisms leading to Figs. 6.2 and 6.5.

6.6 Discussion

We have found that increasing adversarial burden triggers hypersensitivity to 'noise'. This might be expected to cause debilitating arousal 'anxiety attacks' or, in contrast, a switch to pathological desensitization, at both institutional and individual levels of organization. The model tracks this dynamic via the $m = 0, -1$ branches of the Lambert W-function, inherent to the calculation of 'temperature' g for a large class of basic, underlying probability distributions.

Again, Israeli strategic failures in 1973 and 2023 seem to provide canonical examples of the mechanism at the institutional scale, as does the USA's persistent disregard of intelligence regarding the 1950 buildup of a quarter-million strong PLA force at the Yalu river boundary between China and North Korea. More recent USA failures, in Vietnam and Afghanistan, also come to mind.

Psychic numbing at the individual level in the aftermath of severe trauma, as studied by Robert Jay Lifton (1982), has also been well characterized.

On the other hand, hypervigilance is a persistent symptom of post traumatic stress disorder. In this regard, Chiu et al. (2024) write

> According to Ehlers and Clark (2000)'s model, PTSD could be conceptualised as a sense of current threat maintained by maladaptive appraisals of trauma and its sequelae. Elevated anxiety sensitivity may constitute a vulnerability factor that intensifies maladaptive trauma appraisals, in particular those related to anxiety sensations...

We have explored a canonical and highly pathological idiotypic network failure of convergence for cognitive systems under stress and equipartition mechanisms driven to failure by rising 'noise'. Applications range across the physiological, individual, social, institutional, machine, and composite scales and levels of organization. Most particularly, the current race to develop 'artificial general intelligence' or machine 'superintelligence' will inevitably run headlong into a brick wall analogous to the culture-bound syndromes of psychopathology that relentlessly confront and confine the human mind, in spite of our considerable evolutionary pruning. After all, machine systems, like the human minds that conceive and build them, are quintessential cultural artifacts.

In sum, the 'computational psychiatry' approach outlined here may have quite general utility, particularly if it can generate robust statistical tools for the analysis of experimental and observational data, the only source of new knowledge, as opposed to model-generated speculation.

Appendix: Immune and HPA Axis Pathologies

Summaries provided by the Perplexity AI system.

Immune Cognition Under Stress

...Persistent or irrational arousal of the immune system produces systemic states that are functional analogs to psychic numbing and hypervigilance, consistent with Irun Cohen's and Henri Atlan's view of the immune system as a cognitive, self-organizing, meaning-generating process rather than a mere reflexive defense system [1–4].

Immune Cognition and Regulatory Meaning

Cohen's 'immunological homunculus' framework treats immune identity as a dynamic network of self-representations; chronic activation distorts this representational stability much the way trauma distorts neural representations of threat and safety. Atlan and Cohen describe immune information as self-organized meaning emerging through recursive feedbacks of recognition and tolerance. When this feedback loop is destabilized—through persistent antigenic 'false alarms' such as chronic infection, stress, or self-reactivity—the system's interpretive flexibility collapses, producing maladaptive attractors analogous to pathological vigilance or detachment [3, 4].

Immune Hypervigilance: Autoimmunity

Chronic immune stimulation produces what immunologists call immune hyperreactivity or autoimmune priming. Physiological autoimmunity, according to Cohen, normally maintains body integrity by ongoing low-level recognition of self-antigens. When context signals misfire persistently, the immune system over interprets benign internal or external cues as threats—comparable to PTSD hypervigilance. Overproduction of cytokines, chronic inflammation, and the recruitment of monocytes and effector T cells (as seen in long-COVID and autoimmune conditions) represent the immunological analogue of perceptual overarousal [5–7].

Immune Numbing: Exhaustion and Tolerance

Conversely, chronic exposure to antigenic stimulation leads to T cell exhaustion, a well-characterized state where effector cells lose responsiveness despite ongoing threat signals. This blunted responsiveness preserves internal stability at the cost of vigilance—functionally equivalent to psychic numbing or learned helplessness. Such 'immunological burnout' arises precisely because the system's cognitive mechanism of meaning attribution (deciding whether a signal 'matters') becomes saturated [8–10].

Integrative Perspective

From a cognitive-systems viewpoint, persistent irritational arousal drives the immune system into two opposite attractors:

- **Hypervigilance (autoimmunity)**: excessive meaning assigned to neutral cues.
- **Numbing (exhaustion/tolerance)**: collapse of meaning and attenuated response.

In both cases, the hallmark is a breakdown of contextual discernment—the cognitive essence of immune function in Cohen's model. Thus, just as trauma dysregulates neural processing of salience and safety, chronic inappropriate stimulation dysregulates immune cognition, impairing its ability to distinguish 'signal from noise' in the body's semantic field.

Appendix References

1. https://pubmed.ncbi.nlm.nih.gov/1476598/
2. https://plato.stanford.edu/archives/spr2025/entries/immunology/
3. https://pubmed.ncbi.nlm.nih.gov/9678751/
4. https://www.semanticscholar.org/paper/Immune-information,-self-organization-and-meaning-Atlan-Cohen/127892c505689c469c25566bfb715e234ccbb6c2
5. https://www.frontiersin.org/journals/neuroscience/articles/10.3389/fnins.2017.00582/full
6. https://www.brainfacts/org/diseases-and-disorders/infectious-disease/2025/lost-in-the-fog-how-viruses-impact-our-brain-function-091825
7. https://bbrfoundation.org/content/how-immune-activation-may-alter-brain-and-cause-depression-related-behavior-during-chronic
8. https://www.frontiersin.org/journals/immunology/articles/10.3389/fimmu.2025.1662145/pdf
9. https://www.annualreviews.org/content/journals/10.1146/annurev-immunol-090222-110914
10. https://www.frontiersin.org/journals/immunology/articles/10.3389/fimmu.2025.1700039/full

HPA Axis Dysfunction

Persistent dysfunctions of the HPA (hypothalamic-pituitary-adrenal) axis under PTSD can be interpreted through the above two-phase theory of bounded rationality, which models stress-induced cognitive pathology as alternating regimes of hypervigilance and psychic numbing. This framework, rooted in information and control theory, elaborates that real-world cognitive systems must continuously allocate limited resources (rate of internal communication, external information intake, and metabolic supply) to maintain adaptive control and minimize 'distortion' in signal processing. Under increasing stochastic noise or adversarial environmental burden, such systems undergo symmetry-breaking phase transitions—mathematically captured through bifurcation into two distinct regimes, aligned with the dynamics seen in PTSD: sustained hyperarousal (hypervigilance) and pathological downregulation (psychic numbing).

Rate Distortion Control and Phase Transitions

The two-phase model formalizes cognitive dynamics using a multi-dimensional version of rate distortion control, wherein the system operates with limited channel capacity amid rising noise and resource constraint. When persistent stress pushes the system past a critical threshold, the model predicts abrupt switching between two dynamic states:

- An 'on' regime (m = 0 branch of the Lambert W function), expressing hypervigilance: sustained elevated response, rapid readiness, and persistent arousal, characteristic of HPA axis hyperactivity in PTSD.
- An 'off' regime (m = 1 branch), corresponding to psychic numbing: functional desensitization, emotional flattening, and possibly HPA axis hypoactivity.

These transitions are interpreted as groupoid symmetry-breakings, with punctuated shifts between attractor states (hyper- and hypo-responsiveness).

Analogy to Immune System Dysfunctions

The previous appendix section draws a parallel between immune system pathologies and cognitive-phase transitions: chronic immune stimulation (false alarms) leads to either hypervigilance (autoimmunity) or exhaustion (immune numbing/tolerance). For the HPA axis, persistent trauma exposure similarly destabilizes regulatory set points, leading to a loss of contextual discernment. The system fluctuates between pathologically heightened output (e.g.,glucocorticoid overproduction and excessive stress hormones) and blunted, refractory states (diminished responsiveness, 'numbing'). In both physiological systems, chronic noise disrupts the feedback loops that maintain adaptive balance, inhibiting the capacity to correctly distinguish signal from noise.

Application to HPA Axis Pathology in PTSD

Applying the model to the HPA axis under PTSD stress:

- **Hypervigilance phase**: The HPA system is locked into an 'always on' alert state, analogously consuming metabolic and informational resources at excessive rates. Cortisol and related stress hormones are persistently overproduced, matching the 'on' regime (hyperarousal). Negative feedback is degraded, so even minor cues perpetuate stress signaling.
- **Numbing phase**: Having exhausted regulatory capacity, the system shifts to an 'off' state—analogous to T cell exhaustion in immune burnout—where reactivity to both intrinsic and extrinsic signals is diminished. This can present as flattened cortisol response, emotional detachment, and impaired adaptation.

Both states arise as stable equilibria in an underlying groupoid symmetry landscape, with transitions triggered by resource depletion, chronic overdrive, or crisis-level noise—the system 'chooses' one pathological attractor or oscillates between them.

Integrative Perspective

The key insight is that, just as immune cognition can be driven into attractor states of autoimmunity (hypervigilance) or tolerance (numbing) by persistent irritation, the HPA axis under PTSD traverses analogous phases. When subjected to repeated trauma or environmental 'fog,' the system's bounded rationality—its limited capacity to integrate, process, and regulate stress signals—results in punctuated, nonlinear pathology. Thus, the two-phase theory provides a unifying model for interpreting persistent pathologies of the HPA axis under chronic PTSD as repeated phase transitions between hyperarousal and psychic numbing, driven by fundamental constraints on cognitive and regulatory resources.

References

Atlan, H., Cohen, I.: Immune information, self-organization, and meaning. Int. Immunol. **10**, 711717 (1998)

Bartolucci, S., Mozeika, A., Annibale, A.: The role of idiotypic interactions in the adaptive immune system: a belief-propagation approach. J. Stat. Mech.: Theory Exp. **2016**(8), 083402 (2016). https://doi.org/10.1088/1742-5468/2016/08/083402

Bennett, C.H.: The thermodynamics of computation. Int. J. Theor. Phys. **21**, 905–940 (1982)

Brown, R.: From groups to groupoids: a brief survey. Bull. Lond. Math. Soc. **19**, 113–134 (1987)

Chiu, H., Low, D., Chan, A., Meiser-Stedman, R.: Relationship between anxiety sensitivity and post-traumatic stress symptoms in trauma-exposed adults: a meta-analysis. J. Anxiety Disord. **103**, 102857 (2024). https://doi.org/10.1016/j.janxdis.2024.102857

Cover, T., Thomas, J.: Elements of Information Theory, 2nd edn. Wiley, New York (2006)

De Boer, R., Hogeweg, P.: Unreasonable implications of reasonable idiotypic network assumptions. J. Math. Biol. **51**, 381–408 (1989)

de Groot, S., Mazur, P.: Non-equilibrium Thermodynamics. Dover, New York (1984)

Derman, E., Miller, M., Park, D.: The Volatility Smile. Wiley, New York (2016)

Dretske, F.: The explanatory role of information. Philos. Trans. Roy. Soc. A **349**, 5970 (1994)

Dreyfus, H.: What Computers Can't Do: Of Artificial Reason. Harper and Row, New York (1972)

Ehlers, A., Clark, D.: A cognitive model of posttraumatic stress disorder. Behav. Res. Ther. **38**, 319345 (2000)

Feynman, R.: Lectures on Computation. Westview Press, New York (2000)

Fjelland, R.: Why general artificial intelligence will not be realized. Hum. Soc. Sci. Commun. (2020). https://doi.org/10.1057/s41599-020-0494-4

Genewein, T., Leibfried, F., Grau-Moya, J., Braun, D.: Bounded rationality, abstraction, and hierarchical decision-making: an information-theoretic optimality principle. Front. Robot. AI (2015). https://doi.org/10.3389/frobt.2015.00027

Jerne, N.: Towards a network theory of the immune system. Annu. Immunol. **125C**, 373–389 (1974)

Khinchin, A.: The Mathematical Foundations of Information Theory. Dover, New York (1957)

Korem, N., Duek, O., Spiller, T., Ben-Zion, Z., Levy, I., Harpaz-Rotem, I.: Emotional state transitions in trauma-exposed individuals with and without posttraumatic stress disorder. JAMA Netw. Open **7**(4), e246813 (2024). https://doi.org/10.1001/jamanetworkopen.2024.6813

Laidler, K.: Chemical Kinetics, 3rd edn. Harper and Row, New York (1987)

Landau, L., Lifshitz, E.: Statistical Physics, 3rd edn. Part 1. Elsevier, New York (2007)

Lifton, R.J.: Beyond psychic numbing: a call to awareness. Am. J. Orthopsychiatry **52**, 619629 (1982). https://doi.org/10.1111/j.1939-0025.1982.tb01451.x

MacKay, D.: Psychophysics of perceived intensity: a theoretical basis for Fechner's and Stevens' laws. Science **139**, 1213–1216 (1963)

Maturana, H., Varela, F.: Autopoiesis and Cognition: The Realization of the Living. Reidel, Boston (1980)

Nair, G., Fagnani, F., Zampieri, S., Evans, R.: Feedback control under data rate constraints: an overview. Proc. IEEE **95**, 108138 (2007)

Ortega, P., Braun, D.: Thermodynamics as a theory of decision-making with information processing costs. Proc. Roy. Soc. A (2013). https://doi.org/10.1098/rspa.2012.0683

Pettini, M.: Geometry and Topology in Hamiltonian Dynamics and Statistical Mechanics. Springer, New York (2007)

Skogbrott Birkeland, M., Sundnes, J.: Advancing the understanding and treatment of post-traumatic stress disorder with computational modelling. Eur. J. Psychotraumatol. (2024). https://doi.org/10.1080/20008066.2024.2360814

Taleb, N.N. (ed.): Statistical Consequences of Fat Tails: Real World Preasymptotics, Epistemology, and Applications. STEM Academic Press (2020)

Wallace, R., Fricchione, G.: Stress-induced failure of embodied cognition: a general model. BioSystems **239**, 105193 (2024). https://www.sciencedirect.com/science/article/abs/pii/S0303264724000789

Wallace, R.: Computational Psychiatry: A Systems Biology Approach to the Epigenetics of Mental Disorders. Springer, New York (2017)

Wallace, R.: Cognitive Dynamics on Clausewitz Landscapes: The Control and Directed Evolution of Organized Conflict. Springer, New York (2020)

Wallace, R.: How AI founders on adversarial landscapes of fog and friction. J. Defense Model. Simul. (2021). https://doi.org/10.1177/1548512920962227

Wallace, R.: Consciousness, Cognition, and Crosstalk: The Evolutionary Exaptation of Nonergodic Groupoid Symmetry-Breaking. Springer, New York (2022)

Wallace, R.: Deception and Delay in Organized Conflict: Essays on the Mathematical Theory of Maskirovka. Springer, New York (2022)

Wallace, R.: Mathematical Essays on Embodied Cognition: Insights from Information and Control Theories. Springer, New York (2025)

Wallace, R.: Hallucination and Panic in Autonomous Systems: Paradigms and Applications. Springer, New York (2025)

Weinstein, A.: Groupoids: unifying internal and external symmetry. Not. Am. Math. Assoc. **43**, 744–752 (1996)

Wikipedia: https://en.wikipedia.org/wiki/Weber%E2%80%93Fechner_law (2025)

The manufacturer's authorised representative in the EU is Springer Nature Customer Service Centre GmbH, Europaplatz 3, 69115 Heidelberg, Germany. If you have any concerns regarding our products, please contact ProductSafety@springernature.com

Printed and bound by CPI Group (UK) Ltd, Croydon, CR0 4YY
05/10/2026
02228059-0006